Mobile Agents

Mobile Agents

William R. Cockayne
Michael Zyda

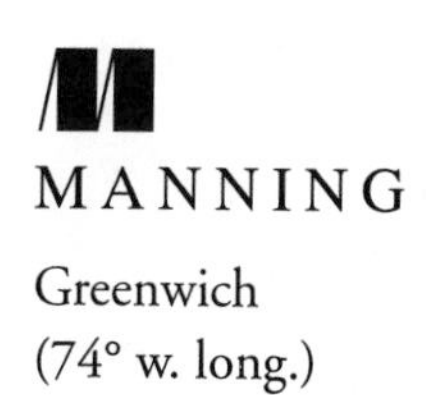

Manning
Greenwich
(74° w. long.)

For electronic browsing and ordering of this book, see http://www.browsebooks.com

The publisher offers discounts on this book when ordered in quantity. For more information, please contact:

Special Sales Department
Manning Publications Co.
3 Lewis Street
Greenwich, CT 06830

Fax: (203) 661-9018
email: orders@manning.com

Library of Congress Cataloging-in-Publication Data
Cockayne, William R.
Mobile Agents /William R. Cockayne, Michael Zyda.
p. cm.
Includes indexes.
ISBN 1-884777-36-8
1. Intelligent agents (Computer software). 2. Internet (Computer network).
I. Zyda, Michael. II. Title.
QA76.76.I58C63 1997
006.3—dc21 97-1595
CIP

Manning Publications Co.
3 Lewis Street
Greenwich, CT 06830

Copyeditor: Elizabeth Martin
Typesetter: Syd Brown
Cover designer: Leslie Haimes

Printed in the United States of America
1 2 3 4 5 6 7 8 9 10 – CR – 00 99 98 97

PART III A Java-based agent system

contents

To John and Betty

introduction

The main incentive to write this book, aimed at readers who want to understand and develop mobile agents for personal or professional use, is the lack of technical agent resources in the popular press. Many of the agent systems presented in this book have never been in print before (except for conference proceedings or journals), making this the first time the nonacademic reader can access this work.

There is a huge amount of literature covering all types of computer agents and the changes that they will effect. Surveying the whole field would overstretch my competence and exhaust your patience; nor will we present a catalog of how to pick the best agent for your business and make millions from it.

What this book will provide is an in-depth look at the world of agents and the promise they hold. This promise can be seen in all of the mobile agent systems that have been developed—the book explains a few of the more dominant systems and how they work, and how to write agents and work with them. Although not a prescription, this book provides the high-level background, in-depth fundamentals, and solid code examples to help you put agents to work in your company.

How to read this book

This book presents a collection of mobile agent systems, each distinct in design and implementation, but similar in goal.

In this book:

- We'll explain what agents are, how they operate, and what they can provide.
- We'll help you better understand the promise the agent technologies presented here provide over today's agents.
- We'll provide an in-depth understanding of the problems that agents can solve, and how these problems have historically been addressed using computer technology.

- We'll explain the technical issues driving the development of agent systems.
- We'll deliver a variety of mobile agent systems which can be used to solve today's problems, as well as address those of the future. Each of these systems is presented so that you understand the developer's goals.
- We'll give you source code examples with analyses that explain the semantics of the agents and how they operate.
- The enclosed CD-ROM provides all of the source code, binaries, and examples for each of the agent systems presented. Additional material on each of the systems, much of it developed by the design teams themselves, is included on the CD-ROM for you to delve further into the areas that interest you the most.
- The CD-ROM also provides pointers to additional reading, other sources of agent information, additional groups working on these problems, and the articles and people driving this work into the future.

The book is presented in three sections. Chapters one and two present the exciting world of mobile agents, some of the underlying technologies, how these agents will be used to solve everyday computer problems, and visions of where agents will fit in our future networks. This first section will be of interest to the reader with little background in this area.

Part II (chapters three, four, and five) presents comprehensive mobile agent systems not developed exclusively with Java. Telescript, Ara, and Agent Tcl are mobile agent systems which have been developed to address some of the issues discussed earlier in the book. Each chapter presents a complete Internet agent system and is co-authored by the main designer of the system. Each chapter explains why the system was developed, some of the major features of the system, the semantics of writing agents for that system, and an example to help the reader get a better grasp of the system. The chapters are written so the reader can develop complete Internet agents when he or she is done reading it.

Part III is comprised of one main chapter and a number of sections presenting IBM's Aglets Workbench, a system designed around the Java language. At the time this book is being written, Aglet Workbench is still beta software. In addition, Java itself is still being developed. Furthermore, the development of a mobile agent system in Java isn't fully understood.

The CD-ROM can be considered additional chapters, making it an integral part of this book. It is important to understand that the CD-ROM supplements the information presented in the book while also being an excellent resource on its own. Although the CD-ROM is physically a separate medium, and is discussed only peripherally in the chapters, please read the following "How to use the CD-ROM" section.

How to use the CD-ROM

The CD-ROM covers a vast amount of information which is not included in the text. In particular, many of the agent systems have additional documentation which can be found only on the CD-ROM.

The CD-ROM also contains the complete software needed to develop and field Internet agents. It also contains references to other agent systems which have been developed, agent resources on the Internet, and tips on obtaining software updates that are developed for the included agent systems.

To make the information on the CD-ROM easier to navigate, the documentation is in `.html` pages which are viewable with any Web browser on all platforms. To begin playing with the agent systems, load the CD-ROM into the computer, open the Web browser on the computer, and open the file `welcome.html`. Feel free to look around.

acknowledgments

I would particularly like to thank:

All of the people who designed the systems in this book and offered the original material for presentation: James White, Holger Peine, and Robert Gray.

Marjan Bace, who thought that a book on mobile agents would be a wonderful addition to the market, and the rest of the staff at Manning Publications who put up with my naivete about the intriguing work of book publishing.

Mike Stegman, who provided the impetus (and text) to tie together many of the grand visions for the book and without whom this book would not have been completed.

Apple Computer, Inc., and my trustworthy PowerBook 540c. People still fail to realize how much a computer can do to make you forget about it completely and allow you to focus on what you really want to do.

My parents, John and Betty.

Bill Cockayne `<cockayne@acm.org>`

PART I

The world of mobile agents

This section presents a general overview of the ideas and technologies behind mobile agents. It gives the reader examples that develop and explain what agents are, what skills they possess, how people are using them today, and where they are headed in the future.

chapter 1

Agents for the Internet

> "A slow sort of country!" said the Queen. "Now, here, you see, it takes all the running you can do, to keep in the same place. If you want to get somewhere else, you must run at least twice as fast as that."
>
> Lewis Carroll, in *Through the Looking Glass*

1.1 Introducing Internet agents

Soon, Alice will have another choice: To stand still and let someone else run for her. The idea of having a personal runner chasing down information has its appeal to those of us who have been blown sideways by yet another shockwave from this century's information explosions. Imagine Alice with a cadre of runners, all pursuing her interests all of the time and leaving her with nothing to do but to stand and wait for their return. As long as we're imagining, let's include runners with the capacity to focus their efforts based on an analysis of Alice's own likes and dislikes. Finally, let's use the Internet as the course to traverse, and we imagined one manifestation of the Internet agent—software that acts on your behalf.

Already other agents have found their way into our daily lives, reminding us of appointments, pointing out spelling errors, or periodically dialing in for our email. While some Internet agents can perform complex information gathering strategies autonomously; others Internet agents can gather discrete information from a limited set of sites, assembling a digest of topics of interest.

Agents, then, perform a service by either being reactive, responding to changes in their environment, or proactive, seeking to fulfill goals. Further, agents can remain stationary by, say, filtering incoming information, or become mobile, searching for specific information across the Internet and retrieving it. As it performs these actions, an agent ideally follows the first principle of good Internet behavior, "Do no harm." Finally, a key element of an agent's behavior, its autonomy, suggests that once goals are established its behavior is guided by its own capacities for action independent of intervention from its user.

1.2 The present, some examples

Even Alice got more help navigating than the average Web surfer gets when he/she first encounters the tangled Web that's been woven across the world. Two or three years ago Lycos and Yahoo! were the most famous locations for beginning any plunge into the

Internet. Now, a count of Internet search engines varies from between one hundred and over five hundred. Some of these engines are simply variations on Lycos or Yahoo!, while others are highly specialized. For example, you can use the Hypertext Webster Interface to search for definitions in Webster's Dictionary[1] or SIFT, the Stanford Information Filtering Tool, developed by Tak Yan at Stanford University, which includes two services, one for computer science technical reports and one for USENET news articles.[2]

Yahoo! presents its information hierarchically: users drill down through the layers of categories, refining the search for information with each selection. The success of this system depends upon the accuracy and judgment of both the people who submit their sites and the Yahoo! team that categorizes them. Additionally, Yahoo! uses its own cataloging software to organize itself. This mix of trained and untrained catalogers complicates the reliability of consistent, accurate listings. To supplement its own hierarchical scheme, Yahoo! also offers the option to search its database using a simple keyword search interface. This option also helps to overcome the silent vagaries of inaccurate cataloging to which Yahoo! is prone.

Lycos introduced brute force indexing of the Internet by using a program often referred to as a spider to search the World Wide Web every day and update its database of indexed sites. In the beginning, Lycos provided a simple search interface; however, when it moved from a university-sponsored research project to a commercial venture, it added other refinements such as a hierarchically organized subject guide and several prepared searches such as an interface for finding stock prices. Similarly, AltaVista, a search engine developed by Digital Equipment Corporation, also catalogs a vast quantity of pages and has recently added LiveTopics to assist users with refining their search by providing a way to select and eliminate search topics.

This effort to simplify searches and lower the inevitable frustration prompts sites like Yahoo! and others to encourage users simply to enter key words describing a topic and take a chance. Such an unstructured approach leads to a lot of hits or nothing at all. A dedicated user can find additional options, generally unique to each search engine's protocol, that claim to refine the search even further. Some sites even offer the chance to select the importance of one term in a search over another. In the end, the various search interfaces all attempt to overcome a basic problem: framing an effective and efficient question for a search requires a complex understanding of how knowledge may be structured within the unstructured universe of the Internet.

---

1 http://c.gp.cs.cmu.edu:5013/prog/webster

2 ftp://db.stanford.edu/pub/sift/sift-1.1-netnews.tar.Z

While librarians rely on a highly refined cataloging scheme to locate a book in a library, untangling the Web resembles the enigma of the Gordian knot. There simply are no cataloging rules.

The vast amorphous tangle that Lycos, AltaVista, Excite, Web Crawler, or any of the other search engines must confront suggests the magnitude of difficulty and frustration awaiting anyone hoping to use the Web as an efficient source of information.

A search of the Web for information on Internet agents produces a blizzard of responses that include references to real estate agents, insurance agents, theatrical agents, houses for sale, and so on, often with the most needed pages sorted at the start, then, maddeningly, another burst of relevance appears some twenty to thirty pages of pointless citations later. In short, a simple search is never simple.

Another recent practice with Web page creation has only added to the clutter of pages returned as a response: keywords. In an effort to ensure that search engines generate numerous hits to the same page, Web page creators have begun to frontload the keyword sections of their Web pages, often piling on ten, twenty, even thirty occurrences of the same word so that a search engine will then place the site higher in its rankings. This practice virtually guarantees that there are pages that will show at the top of completely unrelated searches and guarantees the additional, wasted time to review and page past these repetitions.

As an answer to the incompleteness of any search engine's database, as well as a method for combating excessive hits, meta-search engines have been developed. These engines often perform some preprocessing of keywords before submitting the search to other services. Some of these systems then add postprocessing of the results in an effort to reduce redundancies and to rank order the hits. Some of these meta-search engines include iFind,[3] MetaCrawler,[4] and SavvySearch.[5]

While these meta-search engines seek to cover more of the available indexed sites and often rank order a query's successful hits, they continue to rely on simple brute force to complete their searches.

Into this chaos steps the first important generation of Internet agents. Some offer to reduce the results garbled by inefficient searches and the noise generated by duplicated pages. Others keep users abreast of particular sites, maintaining the latest information from that site by checking it regularly and downloading new information. And still others will remember every page you've visited, indexing and cross referencing matching words for later searches. In short, these agents act to filter, retrieve, or index information.

---

3 http://mf.inference.com/ifind/

4 http://www.metacrawler.cs.washington.edu:8080/

5 http://www.cs.colostate.edu/~dreiling/smartform.html

Some of these agents, available commercially, include SurfBot from Surflogic LLC[6] and WebCompass from Quarterdeck Corporation.[7] Surfbot and WebCompass offer users the chance to use their existing search agents, modify their criteria, and schedule their actions.

Beyond these search engines, a growing group of agents is beginning to edge onto the Internet. Some take the form of scripts for existing technology such as mailers, others are personal implementations of agent technology focused on fulfilling personal needs.

1.2.1 Discussion help, a stationary agent

The problem: To allow topics under discussion throughout a company to be accessible to all employees, using a browser, in an organized fashion.

The solution: An agent that uses a scripting system and integrated database to allow employees to read and post messages to discussion threads.

Dave Winer of UserLand Software[8] has built such a system and provided to the Internet community for free. To handle this on our system we have set up a Web server and a copy of Dave's Frontier scripting environment,[9] also provided to the Internet community for free. Then we installed a BBS system[10] Dave wrote (yes, it's free) and everything was ready to go. Anyone who wants a message posted to the bulletin board goes to the BBS' Web page and logs on using his or her email address. The person who maintains the BBS can set up any number of discussion groups and can manage the messages as the system becomes too large to handle. Since the BBS is on a Web server, the new message is now posted to everyone on the net.

With such an agent in place, the history of a topic's discussion is readily available to all. Such a history also allows a team or a project newcomer the chance to catch up on the discussion's progress without having to find someone to reassemble the thread's developing ideas.

---

6 http://www.surflogic.com/

7 http://arachnid.qdeck.com/qdeck/products/webcompass/

8 http://www.userland.com/

9 http://www.scripting.com/frontier/

10 http://www.scripting.com/frontier/utilities/bbs.html

Although this agent works in a specific environment and relies on a set of scripts to log, post, and maintain the messages, it manages to keep all of the discussions at our site neatly organized. As an elementary agent, ours satisfies our needs and also points toward what will be a developing use for agents.

Just to burst the bubble, this really isn't an agent. Why not? Many of you may have noticed that this system is really just a bulletin board system. BBSs have been around far longer than the Internet and have always been able to do what Winer's system does. The BBS that Winer is distributing is a step toward agent technology compared to the BBSs with which many of us are more familiar. The use of scripts and a general architecture (Frontier) to build this specialized system are features that agents promise. All the BBSs I used in college were specially written programs that ran on one machine and usually didn't allow you to do anything else (like word processing or email) while it was running. Forget it if you wanted any special features or wanted to change the way the BBS worked. The BBS we use is built on top of Frontier, which wasn't built with the intention of creating a BBS for me to use. Someone looked at all the power Frontier has, and came up with something really cool to do with it.

But there are a number of features that the BBS doesn't (and really can't) provide that I want. The BBS and Frontier were designed as status systems: they run on only one type of computer (machines running MacOS), they reside on only one machine, and their behaviors do not change as time goes by.

This BBS goes a long way in showing us how to build something that helps us organize our conversations using a generic architecture. Removing the static-ness of the BBS lets us see what the next generation will provide. We can do this using the mobile agent systems presented in Parts II and III; each is a generic system but with the added ability of movement.

The agent version has the ability to run on a variety of platforms, whether by using an already transportable language like Java (this is what Aglets and the future Ara and Agent Tcl do) or by sitting on top of a Java-like virtual machine (which is what Telescript, Ara, and AgentTcl currently do). So now that the agent can run on different machines, it can move around the Internet. Now the BBS begins to take on a whole new meaning.

What is this future BBS? It could evolve into the next version of a software package like IBM's Lotus Notes, an application that helps people share information. It could be used to help children all over the world collaborate on projects by providing connections to other schools, helping the children and teachers understand the context of the other students, or even create friendships among children who will probably never talk to each other face-to-face.

Imagine this scenario: The agent of a small school in the United States finds out from an agent run by a university that a school outside Moscow is discussing similar issues related to pollution in their communities. The U.S. school's agent contacts the Russian school's agent and begins to coordinate information to show to the teachers of the respective schools.

In the morning each of the teachers will come to school and read about the other school's work, its children, and some of the conversations that have been going on. By that afternoon the school's agents are solving any issues of coordinating each school's BBS. For the rest of the year the children in both countries get to share in the experiences and lives of their schoolmates.

Did I mention that the children and teachers at the two schools don't speak each other's language? The agents found a program at another university that allows the conversations to be translated.

It will happen.

1.2.2 *Classified help, an Internet agent*

The problem: Managing a career as a computer contractor/consultant.

The solution: An agent that automatically checks selected classified categories in the Sunday Washington Post and emails résumés with no intervention required; another agent that checks and saves the total information for selected classified categories; and a third agent, in development, that downloads selected classified sections, and creates a searchable database of job listings.

Charles Crizer (`cfcrizer@dyncon.net`) is a computer contractor who developed these agents to assist him in his search for new jobs. On a typical Sunday the first agent will go to the Washington Post's classified ad site and download all of the jobs filed under selected categories such as technology, programmer, computers, and so forth. Once the agent has done this, it parses the files for email addresses and stores this information along with the date the ad was downloaded. Crizer then activates the agent to search this information and it then emails his résumé to the email addresses that match his criteria. Sometimes, the agent will pump out as many as a thousand résumés in the five minutes its takes to manage this task and it does so with no intervention on Crizer's part other than establishing the date range for the ads the agent should access.

Each Sunday, a second agent downloads the total information from various classified sections and stores them for perusal later. Crizer is developing a variant of this agent which will not only download the classified ads for a section, but will strip out the HTML code, parse each ad, and generate a database that, later, can be searched productively and used to assemble mailing lists for résumés, or a list of available jobs meeting specific criteria.

Just as with the BBS example, this isn't really an agent, but this system also provides some of the features that we want from our mobile agents. One of the major features of this system is that it saves Crizer lots of time while performing necessary but mindless tasks (sending out the résumés).

The mobile agents will allow Crizer to build agents that save him more time while increasing the value of the information that he has to review. Right now the script Crizer has written searches only his local newspaper. With the growth of telecommuting, Crizer may soon be able to answer job ads in different parts of the country by telling the mobile agents to search everywhere for the types of jobs he likes. Since the agents will be looking in different parts of the country, they can wander on the 'Net all week long.

One day, Crizer will wake up to find that one of his agents not only found him the best job he has ever seen, but the agent has already sent his résumé to the company. In this scenario, before Crizer even finishes his first cup of coffee, one of his agents reports back that the company sent an agent requesting his schedule for the day (and was responded to instantaneously by another agent) and that a Mr. Smith would like to call at either 1 or 3 o'clock this afternoon to discuss the position. A job search, complete with interview, and Crizer hasn't even had a chance to butter his toast.

1.3 The future, some examples

The early group of agent-like technologies has responded to the need to overcome the tedium of searching and sorting, always the bane of information whether the form was a paper file and filing cabinets or their electronic parallels. Often they follow rules set up and maintained by their users, and only those rules. They can and still do maintain subscription services for mailing lists and can assemble a personal newspaper each day, but they do not sense changes in your interests nor do they explore beyond their defined rules.

Beyond handling massive information searches off-line, or in the background, the latest agents are focusing on other business-related tasks such as agents that can assist purchasing, finances, travel, and schedules. With them in place a user does not have to scroll through the avalanche of sites a search engine now produces.

1.3.1 Purchasing

Anyone who has ever had to provide three cost comparison prices for each item in a purchase order will immediately recognize the value of an agent that can assemble this information autonomously. The individual purchaser looking for the best price for a single item such as running shoes or memory chips can also benefit from an agent that scours the Internet for the best prices and suppliers.

1.3.2 Finance

A trained agent can monitor a stock profile, calculate its net worth, track market movement, and notify a user of anomalous changes within the market. With this type of an agent running either in the background or off-line, it gathers specific information, analyzes it, and presents the information only when the analysis triggers the need for the user to take action or review. Of course, search agents can also track changes at the sites for companies in which a user holds stock, ensuring the latest information is always available for reading. Finally, agents can watch for new companies as they come on-line and become potential investment opportunities.

1.3.3 Travel

A travel agent would be able to assemble all of the necessary components for a trip including selecting the best flight, hotel, and ground transportation. Additionally, the agent would monitor these components for changes in pricing, etc., that could improve the trip or its cost. In the end, the agent would provide a detailed itinerary, having made and confirmed all of the necessary bookings.

1.3.4 Personal assistant

A personal assistant agent would manage a user's meeting schedule and poll other participants to arrange for a common time and convenient location. In preparation for such a meeting, an agent could assemble and disseminate relevant documents keyed to the meeting's agenda.

1.3.5 Entertainment scheduling, an extended example

The present scenarios for information gathering rely on at least two rather limited technologies—human intervention and blind acquisition. Neither of these types of systems provides for more complex searches where a user may have a conceptual understanding of the information needed, but not the understanding needed to find it. In addition, these systems do not presently store information about other activities in which the user participates on a daily basis, forming a user profile that could assist in streamlining a search by eliminating information unlikely to be of interest.

For example, if the user is interested in finding information on upcoming concerts, the present systems provide a couple of options. The user could use a content word search (AltaVista-type systems) to look for any pages that contain words that may be related to events going on in his area of interest, a highly inefficient way to accomplish such a blind search. Or the user could go to a more categorized system (Yahoo!-type systems) and look for the same type of areas, a slightly more efficient search method. Finally, the user could be lucky enough to have a locally maintained system (Yahoo! Metros) which could maintain information and news of local interest. This last system, however, would still be limited to content word searching or categorized searching: the limiting factors of the first two types of systems.

With Internet agents and the interface to access them, a user would request two tickets to a show for the next weekend. The interface agent could use information gleaned from the knowledge it has accumulated about the user to prep a number of Internet agents. The Internet agents could then be handed information to initiate a search using information such as the following:

- The approximate cost of events the user has previously attended
- The type of previous events the user has attended
- A request to contact the user's friend for helpful information on preferences
- Any scheduling conflicts the user and friend may have
- Arenas the user prefers to attend
- Other information related to preferences or previous situations

Using this profile, each of the Internet agents can then go off to collect information specific to its agenda. In doing so, each can:

- Choose its own path
- Make decisions based on its knowledge
- Keep in contact with the user's other Internet agents to coordinate information
- Contact the interface agent for further user information
- Follow links that may not seem obvious to the user, but are based on information obtained from other users' Internet and interface agents
- Reserve tickets if the events the agent finds are in short-supply
- Build a number of itineraries to present to the user

The time the user would need to build the multiple scenarios that the agents present is not known simply because humans don't have the time or the patience to do this type of extensive work. Agents, on the other hand, have no care for how much time something takes or how seemingly hard it is. The agents' only real limits are the number of machines they can access and the amount of cybercash they possess to perform their duties.

1.4 Lots of simple agents

The promise of autonomous, networked agents is built on the belief that while one agent working alone is good, many working together is even better. In the past, people have attempted to solve complex problems with ever more complex agents. This new idea turns that on its head by leading toward the development of collaborative agents.

Since the Internet agent community is willing to allow agents to work together, developers can enhance the intelligence of such a collective system by adding in human experience to focus its energies. Although these agents promise to free humans from monotonous, time-intensive searches, the presence of a user profile can reshape the expansive, brute force capabilities of a search engine and shape its energies toward more precise and efficient searches.

A recent example of this type of thinking has been presented by Firefly Network, Inc.[11] Firefly is a collaborative filtering system, meaning that it constructs a personality profile or a user from a series of questions and then compares those opinions with other users on the system. Similarities between users are then used to expand the range of music recommendations offered to a user. It also maintains a listing of prototypical users which can focus searches more quickly.

From the beginning, this group of agent proponents, one of the first public spin-offs from Pattie Maes' Agent Lab at the MIT Media Lab, focused on ways to build smart systems out of collections of agents rather than build one huge agent. In one of the early systems at the Media Lab, a mail program watched a user and gathered statistical information about the user's reactions to different messages. The agent built a personal profile of the user, utilizing only statistical understandings of some of the areas that the agent could watch: the mail message's sender, how long before the user deleted a message, and to whom the user forwarded mail. Using simple data on a number of tasks, the agent exhibited remarkable intelligence by further correlating data about the difference;

---

11 http://www.ffly.com/

for example, mail received from a certain mailing list is saved to a certain folder, while mail from the boss is immediately read and forwarded to two friends.

The development team at MIT continued this work by building one of the first real collaborative agent systems. The system revolved around the fact that each agent had strong information about how its user reacted to a variety of mail situations. The researchers advanced the system by allowing agents from a number of different users to get together at night to share information. These meetings were done in a blind manner so no personal information was compromised.

For example, the new user in the company won't have many strong rules for her agent to work with. When her agent meets with more experienced agents, it can ask for any rules they might want to share. A number of agents in the collective know that their users always read mail from the boss immediately. They convey this information to the novice agent, who can then present this to its user the next morning. The user decides whether to accept this rule, and the agent obtains a new rule and some information (again statistical) on what agents in the collective provide rules its user likes. Already Firefly suggests how such searches might occur by presenting users with a personalized system for recommending music.

1.5 Present agents and Java agents

The systems presented above were written by one group or company and will run only on special machine or networks; the agents written in Agent Tcl will only be able to move among Agent Tcl-based servers and use services based on Agent Tcl servers. This is by design. One of the design constraints that affected this decision is the fact that the foregoing systems, and a number of systems in this book, were developed to provide the best possible agents. At the time these systems were devised (back in prehistoric days of the mid 1990s), this meant designing everything in the system; this gave the creators and users the most agent functionality.

Telescript, presented in this book, was the first commercially available mobile agent platform. Telescript, created by General Magic and backed by a number of technology and telecommunications companies, was based on the idea that agents could be used to help users as they tracked down information and to facilitate business transactions. Because Telescript attempted this before the Web really existed, much of the promise of networks had not yet been realized by the consumer. Although a number of factors could be blamed, its main problem was that it was proprietary. As such, the agents could only run on special devices and on special networks. Telescript has since been realigned to work on the Internet.

Java promises to deliver us from proprietary solutions because it runs on all platforms, was designed for mobile code, and was designed for the Internet. The only problem with Java is that it is a young language and is a moving target for people to develop on. Because of this, there are few agent systems developed in Java, and none of these systems provide the robustness and well-thought-out architectures that the complete systems presented above possess.

1.6 How much longer?

How long will it be before Internet agents are pervasive and easy to use? The answer is independent of the technology needed to implement Internet agents. As you will see, each of the systems presented here is stable enough to be deployed today. Indeed, Telescript was actually deployed a number of years ago. Many of the decisions presently driving the development of the Internet are based on the goals and desires of companies fighting to maintain their status in the area of future business. As more people become connected, the value of the Internet to other users and to businesses increases exponentially. In order for Internet agents to be of a real value to users and businesses, groups need to start using Internet agents in realistic ways. However, what will surely bring the widespread use of agents to critical mass will be the nexus of time and information: too much information and not enough time to sort through it all.

At present, some intelligence is embedded in computing machines while most is built into people, and there is almost none in the network itself. Using agents, this new model takes advantage of what computers are really good at: remembering and processing large amounts of information; what people are good at: thinking about trends and higher-level understanding; and what networks are good at: maintaining connections and moving information and knowledge around.

chapter 2

Enabling mobile agents

2.1 *Mobile agent paradigm*

The concept of a mobile agent sprang from a critical examination of how computers have communicated since the late 1970s. Prompted by the difficulty of the Internet's then-current architecture to match the pace of the exponential growth of its users, a new approach was needed that would satisfy two seemingly contradictory needs: increasing the sophistication of the types of communications possible without strangling the available bandwidth of the Internet's weaker components. This chapter sketches the results of that examination and presents the case for mobile agents: a solution that satisfies the basic needs of increased sophistication without a crippling sacrifice in bandwidth.

2.1.1 *Current approach*

The central organizing principle of today's computer communication networks, Remote Procedure Calling (RPC), was conceived in the 1970s and viewed computer-to-computer communication as enabling one computer to call procedures in another. Each message that the network transports either requests or acknowledges a procedure's performance. A request includes data that are the procedure's arguments and the response includes data that are its results. The procedure itself is internal to the computer that performs it.

Two computers whose communication follows the RPC paradigm agree in advance upon the effects of each remotely accessible procedure and the types of its arguments and results (see figure 2.1). Their agreements constitute a protocol.

A user computer with duties for a server to accomplish orchestrates the work with a series of remote procedure calls. Each call involves a request sent from a user to a server and a response sent from the server to the user. For example, to delete all files at least two months old from a file server, a user's computer would have to make two calls: one to get

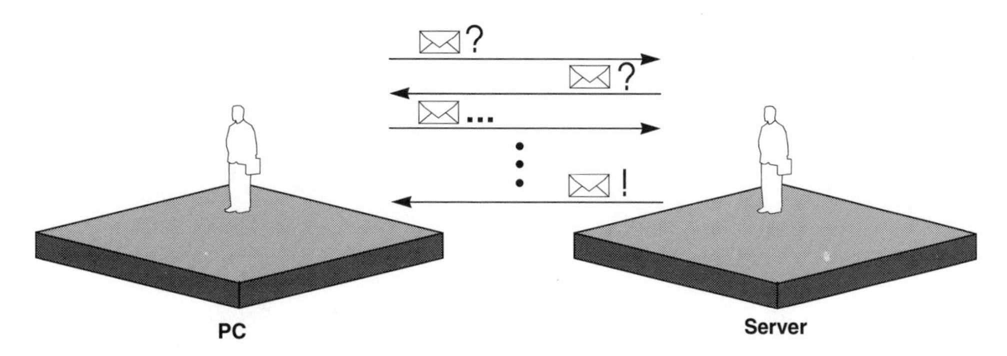

Figure 2.1 The request and response calls

the names and ages of the user's files and a second for each file to be deleted. The analysis that decides which files are old enough to delete is done in the user computer. If it decides to delete n files, the user computer must send or receive a total of $2(n+1)$ messages.

RPC essentially requires that each interaction between the user computer and the server consist of two acts of communication: one to ask the server to perform a procedure and another to acknowledge that the server did so. Because of this, each ongoing series of interactions requires a similarly ongoing series of communications. The clutter in RPC comes from the constant need for reciprocity between the user's and a server.

2.1.2 New approach

An alternative to RPC is remote programming (RP). The RP paradigm views computer-to-computer communication as enabling one computer not only to call procedures in another, but also to supply the procedures to be performed. With RP, each message that the network transports is composed of a procedure that the receiving computer is to perform and the data that are that procedure's arguments. In an important refinement, the sending computer begins or continues a procedure to be performed while the receiving computer continues the procedure using the data sent as the procedure's arguments to define its particular actions, thus defining the procedure's current state (see figure 2.2).

Two computers communicating with the RP paradigm agree in advance upon the instructions that are allowed in a procedure and the types of data that are allowed in its state. Their agreements constitute a language. The language includes instructions that allow the procedure to make decisions, examine, and modify its state, and, importantly, call procedures provided by the receiving computer. Such procedure calls are local rather than remote. The procedure and its state are termed a mobile agent to emphasize that they represent the sending computer even while they reside at and operate in the receiving computer.

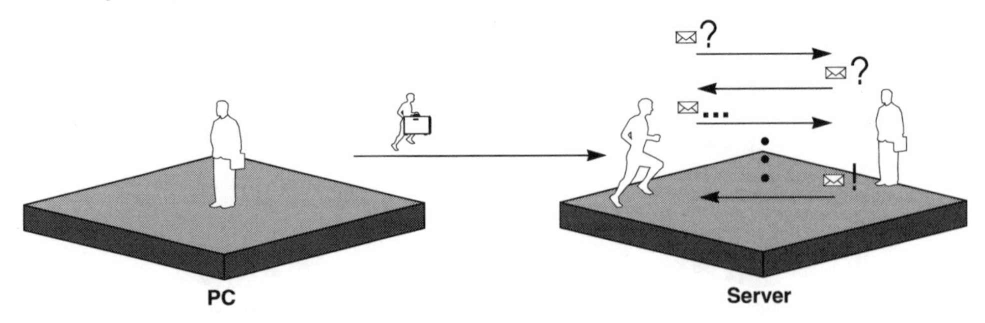

Figure 2.2 Defining instructions

For example, a user computer with work for a server to accomplish sends to the server an agent whose procedure, while there, makes the required requests of the server (e. g., *delete*) based upon its state (e. g., *all files older than two months*). Deleting the files described in the RPC example, no matter how many, requires just the message that transports the agent between computers. The agent, not the user computer, orchestrates the work, deciding *on-site* which files should be deleted.

Using the remote programming paradigm, a user's computer and its server can interact without using the network once the network has transported an agent between them. Thus, an ongoing interaction does not require a series of ongoing communications. This type of interaction appears to achieve the new approach's requirements of increased sophistication without increased bandwidth. Indeed, RP suggests that in some cases, more can happen with less bandwidth.

Remote programming's important advantage over remote procedure calling can be seen from two different perspectives: one, quantitative and tactical; the other, qualitative and strategic.

2.1.3 Tactical Advantage

The immediately apparent tactical advantage of remote programming is performance. When a user computer has work for a server to do, rather than shouting commands across a network, it sends an agent to the server and directs the work locally rather than remotely. The network is called upon to carry fewer messages. The more work to be done, the more messages remote programming avoids.

The performance advantage of remote programming depends in part upon the network: the lower its throughput or availability, or the higher its latency or cost, the greater the advantage. The standard telephone connection presents a greater opportunity for the new paradigm than does an Ethernet network. Today's wireless networks present greater opportunities still. Remote programming is particularly well suited to personal communicators, whose networks are slower and more expensive than those of an enterprise's personal computers as well as personal computers in the home where the telephone line available is largely a voice line.

A home computer or a wireless communicator are examples of a user's computers that are connected to the Internet occasionally rather than permanently. Remote programming allows a user with such a system to delegate an agent to perform a task, or a long sequence of tasks. The system does not need to be connected while the agent carries out its assignment. Thus remote programming allows computers that are connected only occasionally to do things that would be impractical with the current use of remote procedure calling.

2.1.4 Strategic advantage

The strategic advantage of remote programming is customization. Agents encourage the manufacturers of user software to extend the functionality offered by the manufacturers of server software. As noted before, if a file server's software provides one procedure for listing a user's files and another for deleting a file by name, a user can effectively add a procedure that deletes all files of a specified age. This new procedure, which takes the form of an agent, customizes the server for that particular user.

The remote programming paradigm changes not only the division of labor among software manufacturers, but also the ease of installing that software. Unlike the standalone applications that popularized the personal computer, personal communicators will employ communicating applications with components that must reside on servers. The server components of an RPC-based application for a personal communicator must be statically installed by the user. The server components of an RP-based application, on the other hand, are dynamically installed by the application itself running from the user's computer, since each component is an agent.

While the advantage of remote programming is significant in an enterprise network, it becomes a profound advantage for a public network where servers are owned and operated by public service providers. Introducing a new RPC-based application requires a business decision on the part of the service provider. For an RP-based application, all that's required is a buying decision on the part of an individual user. Remote programming thus makes the Internet, like a personal computer, a platform.

2.2 Mobile agent concepts

The first commercial implementation of the mobile agent concept, General Magic's Telescript technology, attempted to allow automated as well as interactive access to a network of computers using mobile agents. The commercial focus of General Magic technology, the electronic marketplace, requires a network that will let providers and consumers of goods and services find one another and transact business electronically. Although the electronic marketplace still does not exist fully, the Internet has already encouraged its beginnings. Telescript's creators envision the electronic marketplace as only a small piece of the agent world that will exist in coming years. There, agents will act on their user's behalf to research information for work, find the best hotel for vacation, provide up-to-the-minute scores of sporting events, or simply send and receive messages between friends.

Since General Magic's realization of the basic concept of agent architecture through Telescript is an easy example to follow, we'll use it to introduce how an agent can work successfully. Telescript implements the following principal systems associated with remote programming: places, agents, travel, meetings, connections, authorities, and permits.

2.2.1 Places

Agent technology models a network of computers, however large, as a collection of places offering a service to the mobile agents that enter. For an agent, a mainframe computer might function as a shopping center housing, for example, a ticket place where agents can purchase tickets to theater and sporting events, a flower place where agents can order flowers, and a directory place where agents can learn about any place in the shopping center. The network might encompass many independently operated shopping centers, as well as many individually operated shops, many of the latter on personal computers (see figure 2.3).

Servers provide some places and user computers provide others. For example, the home place on a user's personal communicator might serve as the point of departure and return for agents that the user sends to servers.

2.2.2 Agents

Communicating applications are modeled as a collection of agents. Each agent occupies a particular place. However, an agent can move from one place to another, thus occupying different places at different times. Agents are independent in that their procedures are performed concurrently.

The typical place is permanently occupied by one, distinguished agent. This stationary agent represents the place and provides its service. For example, the ticketing agent provides information about events and sells tickets to them, the flower agent provides information about floral arrangements and arranges for their delivery, and

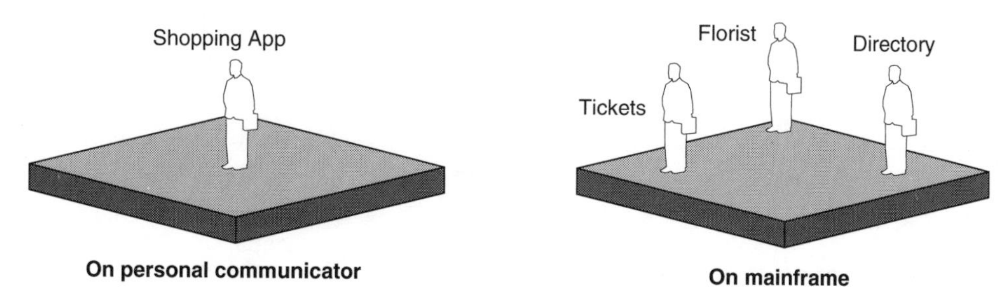

Figure 2.3 A network shopping center

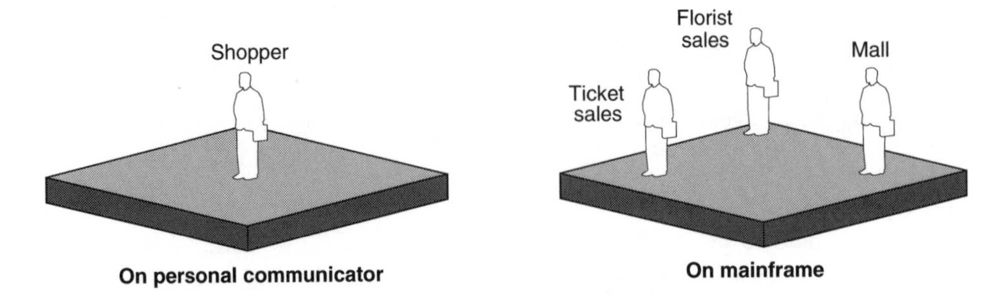

Figure 2.4 Agents mark their places

the directory agent provides information about other places, including how to reach them (see figure 2.4).

2.2.3 *Travel*

Agents are allowed to travel from one place to another, however distant, the hallmark of a remote programming system. Thus, travel allows an agent to obtain a service offered remotely and then return to its starting place. A user's agent, for example, might travel from home to a ticketing place to obtain orchestra seats for a theater show. Later, the agent might travel home to describe to its user the tickets it obtained (see figure 2.5).

Moving software programs between computers using a network has been commonplace for twenty years, or more. In such a case a local area network is employed to download a program from the file server, where it is stored, to a personal computer, where it runs. Contrary to this process, agents move programs while they run, rather than before. A conventional program written, for example, in C or C++ cannot be moved under these conditions because neither its procedure nor its state is portable. An agent can move from place to place throughout the performance of its procedure because the procedure is written in a language designed to permit this movement. The

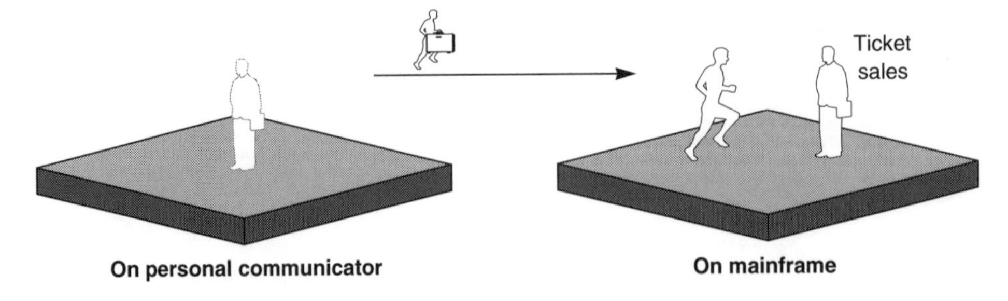

Figure 2.5 An agent carries out a mission

Telescript language lets a computer package an agent, its procedure and its state, so that the agent can be transported to another computer. The agent itself decides when such transportation is required.

To travel from one place to another, an agent executes Telescript's `go` instruction. The instruction requires a ticket, data that specify the agent's destination and the other terms of the trip (for example, the means by which it must be made and the time by which it must be completed). If the trip cannot be made (for example, because the means of travel cannot be provided or the trip takes too long), the `go` instruction fails and the agent handles the exception as it sees fit. However, if the trip succeeds, the agent finds that its next instruction is executed at its destination. Thus, in effect, the language reduces networking to a single instruction.

This `go` instruction lets the agents of different users co-locate themselves so they can interact efficiently. With this capability an agent can arrive at a single place and either converse with a stationary agent, one that resides only in one place, or, if necessary, converse with one or more agents who have also arrived at the same place.

2.2.4 *Meetings*

Two agents are allowed to meet if they are in the same place. A meeting lets agents in the same computer call one another's procedures. The Telescript `meet` instruction used to accomplish this allows the co-located agents of users to exchange information and carry out transactions.

Meetings motivate agents to travel. An agent might travel to a place in a server to meet the stationary agent that provides the service the place offers. An agent in pursuit of theater tickets, for example, may travel to and then meet with a ticket agent. Alternatively, two agents might travel to the same place to meet each other to, say, participate in a venue used for buying and selling used cars (see figure 2.6).

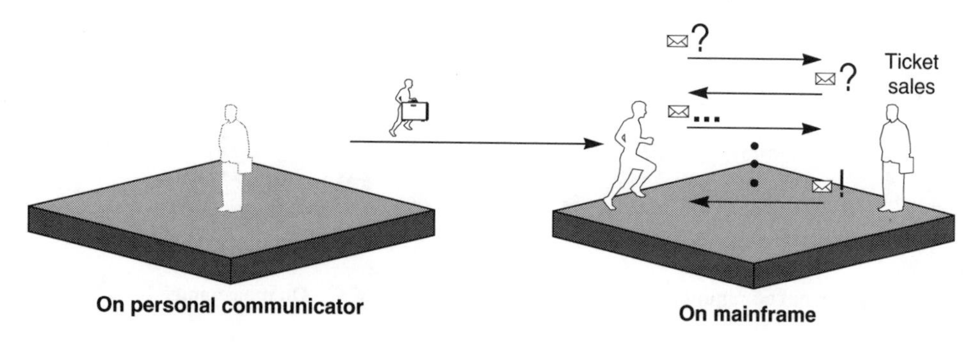

Figure 2.6 Carrying out a `meet` instruction

To meet a co-located agent, an agent executes the "meet" instruction. The instruction requires a petition, data that specify the agent to be met and the other terms of the meeting such as the time by which it must begin. If the meeting cannot be arranged (for example, because the agent to be met declines the meeting or arrives too late), the `meet` instruction fails and the agent handles the exception as it sees fit. However, if the meeting occurs, the two agents are placed in programmatic contact with one another.

2.2.5 Connections

A connection, when two agents in different places communicate, is often made for the benefit of the human users of interactive applications. For example, an agent that travels in search of theater tickets might send to an agent at home a diagram of the theater, showing the seats available. The agent at home might present the floor plan to the user, and, in turn, send the locations of the seats the user selects to the agent on the road (see figure 2.7).

To connect to a distant agent, an agent executes a Telescript `connect` instruction. This instruction requires a target and other data that specify the distant agent, the place where that agent resides, and the other terms of the connection, such as the time by which it must be made and the quality of service it must provide. If the connection cannot be made (for example, because the distant agent declines the connection or is not found in time or the quality of service cannot be provided), the connect instruction fails, and the agent handles the exception as it sees fit. However, if the connection is made, the two agents are granted access to each other (see figure 2.7).

In the agent world, the `connect` instruction allows the agents of users to exchange information at a distance. Sometimes, as in the theater layout phase of the ticketing example, the two agents that make and use the connection are parts of the same communicating application. In such a situation, the protocol that governs the agents' use of the connection is of concern only to that one application's designer. It need not be standardized.

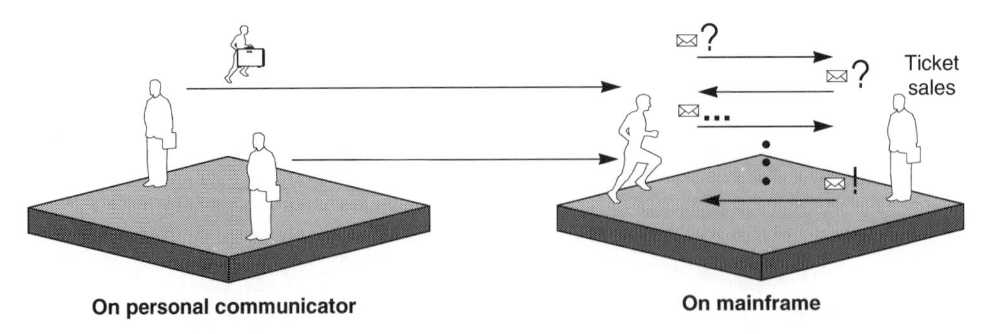

Figure 2.7 Two agents communicate

2.2.6 *Authorities*

This agent system lets one agent or place discern the authority of another. The authority of an agent or place in the electronic world is the individual or organization in the physical world that it represents. Agents and places can discern, but neither withhold nor falsify their authorities, precluding anonymity.

To control access to its files, a file server must know the authority of any procedure that instructs it to list or delete files. This need, important for any network's security, is the same whether the procedure is stationary or mobile. The system verifies the authority of an agent whenever it travels from one region of the network to another. A region is a collection of places provided by computers that are all operated by the same authority. Unless the source region can prove the agent's authority to the satisfaction of the destination region, the agent is denied entry to the latter. In some situations, highly reliable, cryptographic forms of proof may be demanded (see figure 2.8).

To determine an agent's or place's authority, an agent or place executes Telescript's `name` instruction. This instruction is applied to an agent or place within reach for one of the reasons discussed below. The result of the instruction is a telename, data that denote the entity's identity as well as its authority. Identities distinguish agents or places with the same authority.

Authorities let agents and places interact with one another on the strength of their ties to the physical world in three different ways. First, a place can discern the authority of any agent that attempts to enter it, and can arrange to admit only agents of certain authorities. Second, an agent can discern the authority of any place it visits, and can arrange to visit only places of certain authorities. Finally, an agent can discern the authority of any agent with which it meets or to which it connects, and can arrange to meet with or connect to only agents of certain authorities.

The `name` instruction can permit programmatic transactions between agents and places which stand for, say, financial transactions between their authorities. A server agent's authority can bill a user agent's authority for services rendered. In addition, the

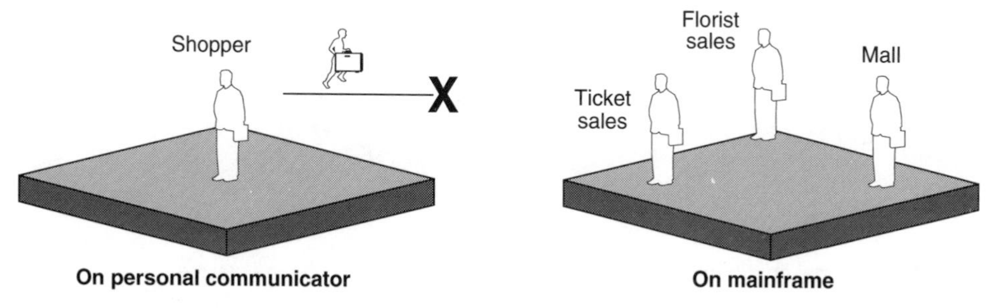

Figure 2.8 Denying entry to an agent

server agent can provide personalized service to the user agent on the basis of its authority, or can deny it service altogether. More fundamentally, the lack of anonymity helps prevent viruses by denying agents that contain the characteristics of a virus.

2.2.7 Permits

Authorities can limit what agents and places can do by assigning permits to them. A permit is data that grants capabilities. An agent or place can discern its capabilities, what it is permitted to do, but cannot increase them.

Permits grant capabilities of two kinds. A permit can grant the right to execute a certain instruction; for example, an agent's permit can give it the right to create other agents. Having done this, the agent must share its allowances of these capabilities and can grant only those capabilities it itself possesses to any agent it creates. An agent or place that tries to exceed any of these qualitative limits is simply prevented from doing so. A permit can also grant the right to use a certain resource in a certain amount. For example, an agent's permit can give it a maximum lifetime in seconds, a maximum size in bytes, or a maximum amount of computation within the limits of its own allowance. An agent or place that tries to exceed one of these quantitative limits is destroyed (see figure 2.9). An agent can even impose temporary permits upon itself. The agent is notified, rather than destroyed, if it violates one of these temporary permits. With this feature of the Telescript language, an agent can recover from its own misprogramming.

To determine either an agent's or a place's permit, an agent or place executes Telescript's `permit` instruction. This instruction is applied to an agent or place within reach for one of the reasons discussed in section 2.2.6.

Permits protect authorities by limiting the effects of errant and malicious agents and places. Such a rogue agent threatens not only its own authority, but also those of the place and region it occupies. For this reason, the technology allows each of these three authorities to assign an agent a permit. The agent can exercise a particular capability only to the extent that all three of its permits grant that capability. Thus, an agent's effective permit is renegotiated whenever the agent travels. To enter another place or

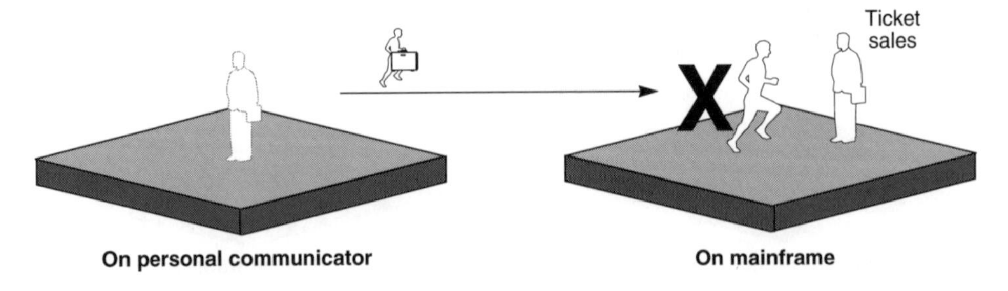

Figure 2.9 Keeping an agent within limits

region the agent must agree to its restrictions. When the agent exits that place or region, its restrictions are lifted, but those of another place or region are imposed.

The `permit` instruction and the capabilities it documents help to guard against the unbridled consumption of resources by ill-programmed or ill-intentioned agents. Such protection is important because agents typically operate unattended in servers rather than in user computers where their misdeeds might be more readily apparent to the human user.

2.2.8 Putting things together

An agent's travel is not restricted to a single round-trip. The power of mobile agents becomes fully apparent when one considers that an agent may travel to several places in succession. Using the basic services of the places it visits, such an agent can provide a higher-level, composite service.

Recalling our ticketing example from above, traveling to the ticket place might be only the first of the agent's responsibilities. The second might be to travel to the flower place and there arrange for a dozen roses to be delivered to the user's companion on the day of the theater event. Note that the agent's interaction with the ticket agent can influence its interaction with the flower agent (see figure 2.10). For example, if instructed to get tickets for any available evening performance, the agent can order flowers for delivery on the day for which it obtains tickets.

This simple example has far-reaching implications. The agent fashions from the concepts of tickets and flowers the concept of special occasions. While the agent does this for the benefit of an individual user in our example, a variation of the example suggests that the agent could also take up residence in a server and offer its special-occasion service to other agents. Thus agents can extend the functionality of the network, conveying the sense that the network is also a platform.

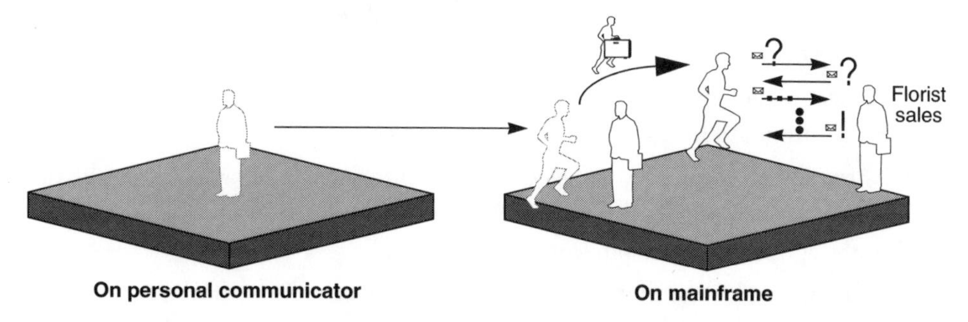

Figure 2.10 Interactions influence outcome

2.3 Mobile agent technology

New communication schemes beget new communication technologies. A technology for mobile agents is software that can ride atop a wide variety of computer and communication hardware, present and future. This technology, which implements the concepts of the previous section and others related to them, has three major components: the language in which agents and places are programmed; an engine, or interpreter, for that language; and communication protocols that allow engines residing on different computers to fulfill the go instruction and exchange agents.

2.3.1 Language

With an agent, programming language developers of communicating applications can define the algorithms that agents follow and the information that agents carry as they travel the Internet. Such a language supplements systems' programming languages such as C and C++. Where entire applications can be written in the agent language, the typical application is written partly in C.

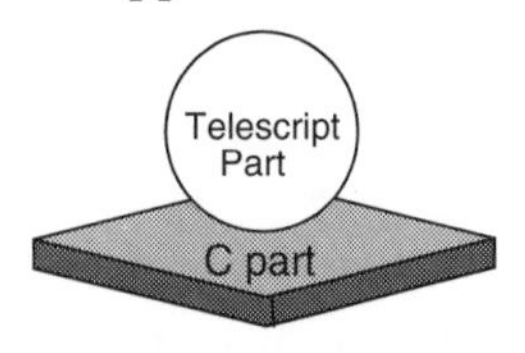

Figure 2.11 The communicating application

The C parts include the stationary software in user computers permitting agents to interact with users, and the stationary software in servers that allows places to interact, for example, with databases. To facilitate the development of communicating applications, the agents and the places to which they are exposed are written in the Telescript language (see figure 2.11) which consists of the following qualities:

- *Complete.* Any algorithm can be expressed in the language. An agent can be programmed to make decisions; to handle exceptional conditions; and to gather, organize, analyze, create, and modify information.
- *Object-oriented.* The programmer defines classes of information, one class inheriting the characteristics of another. Classes of a general nature, such as an Agent class, are predefined by the language. Classes of a specialized nature, such as a shopping agent, are defined by communicating application developers.
- *Dynamic.* An agent can carry an information object from a place in one computer to a place in another. Even if the object's class is unknown at the destination, the object continues to function: its class accompanies it.
- *Persistent.* Wherever it goes, an agent and the information it carries, even the program counter marking its next instruction, are safely stored in nonvolatile memory. Thus the agent persists despite computer failures.
- *Portable and safe.* A computer executes an agent's instructions through a Telescript engine, not directly. An agent can execute in any computer in which an engine is

installed, yet it cannot access directly its processor, memory, file system, or peripheral devices. This helps prevent viruses.

- *Communication-centric.* Certain instructions in the language, several of which have been discussed, allow an agent to carry out complex networking tasks, such as transportation, navigation, authentication, access control, and so on.

2.3.2 Engine

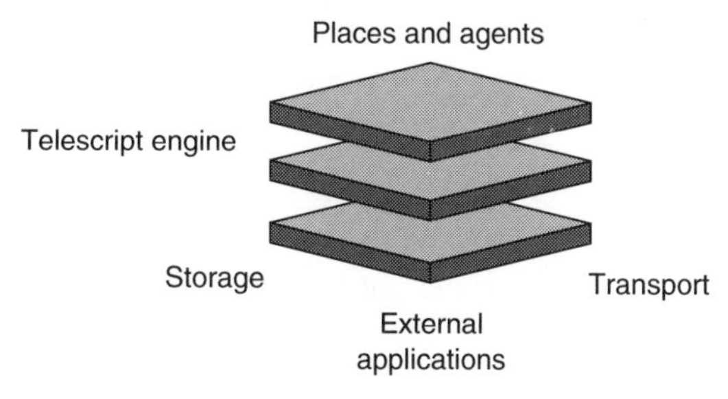

Figure 2.12 A typical engine

The engine is a software program that implements the language by maintaining and executing places within its purview, as well as the agents that occupy those places. An engine in a user computer might house only a few places and agents. The engine in a server might house thousands.

At least conceptually, the engine (see figure 2.12) draws upon the resources of its host computer through three application program interfaces (APIs). A storage API gives the engine access to the nonvolatile memory it requires to preserve places and agents in case of a computer failure. A transport API gives the engine access to the communication media it requires to transport agents to and from other engines. An external applications API allows the parts of an application written in the agent language to interact with those written in C.

2.3.3 Protocols

The protocol suite enables two engines to communicate. Engines communicate in order to transport agents between them in response to the `go` instruction. The protocol suite can operate over a variety of transport networks, including those based on the Internet's TCP/IP protocols, the telephone companies' X.25 interface, or even email (SMTP).

The protocols operate at two levels. The lower level governs the transport of agents; the higher, their encoding and decoding. Loosely speaking, the higher-level protocol occupies the presentation and application layers of the seven-layer Open Systems Interconnection (OSI) model (see figure 2.12).

The encoding rules specify how an engine encodes an agent—its procedure and its state—as binary data and sometimes omits portions of it to optimize performance. Although engines are free to maintain agents in different formats for execution, they must employ a standard format for transport.

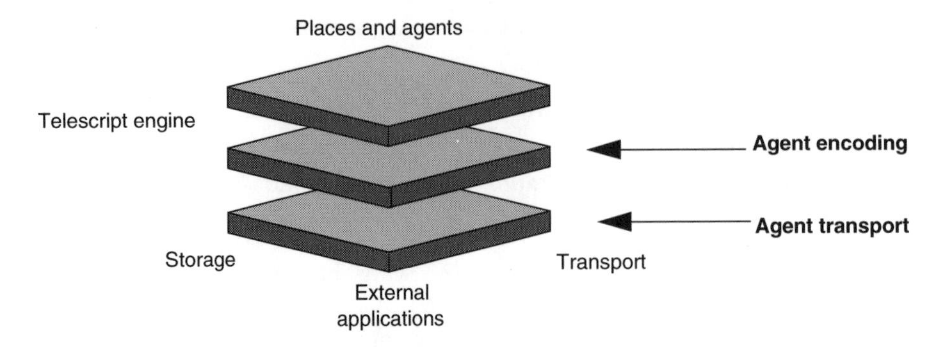

Figure 2.13 The two protocol levels

The platform interconnect protocol specifies how two engines first authenticate one another using, for example, public key cryptography, and then transfer an agent's encoding from one to the other. The protocol is a thin veneer of functionality over that of the underlying transport network.

2.3.4 Legal and ethical considerations

In addition to the technical issues that still face the development of pervasive Internet agent systems, a large number of legal and ethical issues have yet to be adequately explored by the computer community. Most of the discussion of these issues has been in legal and philosophical arenas whose members are often the antithesis of the technical communities. Topics discussed include:

- *Authentication.* This is the major issue behind the use of Internet agents in an open network environment. Just as authentication, verifying the truth of any encounter, plays a major part of people's real world lives, so will it become a major component of agents' lives.
- *Secrecy.* The need to maintain the proprietary information that the continuing financial success an enterprise rests upon will escalate, given an agent's ability to reside and perform operations independent of the server's inherent capabilities. Although the use of a scheme that manages an agent's authority and permission to perform only certain tasks is a response to the need for security, unease, along with the sophistication of agents, is bound to grow.
- *Privacy.* Privacy is oftentimes more important to individuals than it is to companies and governments which often seek to expand their influence while maintaining the proprietary information that hones their competitive edge. An individual's privacy will run the same risks as enterprise secrets; however, individuals may not

be as cautious or protective of their privacy, either from lack of diligence or awareness of the issue.

- *Responsibility.* Ideally, agents must be responsible citizens in terms of how they act when entering a host computer, how they act towards other agents, and how they use their information to deliver the best results to their owner. Properly, issues related to how agents operate fall more in the realm of the developers as a function of a responsible design process. Reading the following chapters on the development of the Internet agent systems will make it obvious that the authors of these systems have spent an extensive development time thinking about and implementing ways in which agents can be kept from harming host systems—and each other.

2.3.5 Why bother?

Although the successful development of useful agents has moved only beyond the embryonic stage, at least three issues specific to agents make them a particularly appealing technology: mobility, adaptivity, and collaboration. Using these three elements developers can build agents of increasing sophistication and autonomy. With the addition of a personal profile of preferences and behaviors as a component of their capabilities, an agent becomes a dynamically responsive expert system, one that learns and changes its behavior synonymously with the maturing preferences of the user on whose behalf it acts. When collaborating with others, an agent can achieve an improved intelligence that absorbs the experiences of all around it.

Mobility gives agents the ability to move around in the agent world to help people solve their problems. No longer limited to a single pathway or restricted to only one action, agents can be sent to explore and establish new and possibly more helpful contacts than a user might, given the time constraints involved. It should be equally clear that the business world could not operate very efficiently if everyone had to stay in their personal office and no one could visit anyone else.

Adaptivity endows an agent with the capacity, when faced with an unknown problem, to discover a new approach or solution. In order to operate effectively, agents must possess a minimal ability to adapt to their environments.

Collaboration drives creative synergy within the business world, making the whole greater than the sum of its parts. Similarly, collaborative agents can share information and join to solve problems. Currently, a number of development groups are exploring this topic independent of the work devoted to simple Internet agents.

2.3.6 Potential roadblocks

Although agents will become increasingly popular, a number of issues could hinder this growth. The most pervasive issue that will slow the growth and acceptance of the agent world are legacy computer systems. Legacy systems, often used to describe the computers from the '60s and '70s still in use today, continue to perform important operations for the companies using them. As the next century comes of age, the machines from the '80s and the machines that are being used today will themselves become legacy systems, exhibiting the same problems that legacy mainframe systems present to us today.

These systems are the lands that time forgot. Limited hardware, limited operating systems based on outdated concepts, green-screen interfaces, and a host of other issues all compound the fact that these systems not only won't, but cannot catch up to the real world. Just as these issues would complicate the acceptance of any new technology, agents will be particularly hard hit. While many new technologies require changes to only one or two parts of a computer and the adoption by only a small group of machines or users for their success, an agent needs a larger number of technical changes and requires mass distribution of its systems to provide real solutions.

Another stumbling block is the inherent caution of those in information services to any technology which threatens a system's security. IS personnel also worry about the uniformity of enterprise standards and practices, as well as the coherent information management of information, and the information's consistency and accuracy.

2.4 Overview of agent systems included in this book

Each of the following chapters focuses on a single mobile agent system. Each chapter was written by the person who designed and implemented the Internet agent system. These authors explain the goals of their projects, the way their system works, and give some examples of agents in their systems.

Each of the systems under study here was designed for one major application area that the others don't focus on. While the other systems may include every area, the focus on one particular solution has greatly affected the design decisions that were made by each of the authors. While the similarity between each of the agent systems may seem to be the dominant message, look for the small differences in the way each author presents his systems and solutions. It is these differences that really define the flavor of the system and what it can be used for.

These differences are what can also help you to decide which system you want to start with first or which one you want to use to solve your problem. When designing your agent network, you will have some idea of what you want to do. Keep your agents' goals in mind as you read about each of the agent systems in order to discover which one best fits your needs. Additionally, you may want to use the following overviews to pick a particular agent system to start with, based on the ideas you already have.

2.4.1 Agents for remote access (Ara)

The basic idea behind the Agents for Remote Access (Ara) system is to develop a software platform for agents able to move freely and easily when they decide to. They can move without interfering with their execution, utilizing various existing programming languages and even existing programs, independent of the operating systems of the participating machines. The Ara system is in active development. The basic system is sufficiently developed for useful applications; many of the more advanced features are not yet implemented and some are still in the process of definition.

The initial section of the chapter introduces the basic concepts of Ara, such as languages, agents, and mobility. The second section demonstrates how common problems of networked computing can be solved using these concepts. This is followed by a section explaining the individual features and facilities of Ara as they are presented to the programmer. These are put to use in an example for searching the World Wide Web. The subsequent section discusses selected aspects of the Ara system architecture to deepen the understanding of the system's capabilities and shows how to extend it with other programming languages. The chapter concludes with a discussion of different approaches to mobile agent systems and future developments of Ara.

2.4.2 Agent Tcl

Agent Tcl was developed by Robert S. Gray and colleagues at the Dartmouth College Computer Science Department. Agent Tcl is based on the Tool Command Language (Tcl) developed at Sun Microsystems and which is now being furthered by Sun as a complement to its Java programming language.

Agent Tcl was developed with four goals in mind:

- Reduce migration to a single instruction, `agent_jump`, and allow this instruction to occur at arbitrary points
- Provide flexible and low-level communication mechanisms

- Provide a high-level scripting language as the main agent language
- Provide effective security.

The software needed to run Agent Tcl is included on the CD-ROM. Any reader interested in further developing agents in Agent Tcl should visit the Tcl Web site looking for new features or updates (the URL is included on the CD-ROM).

The examples present an architecture for implementing a variety of remote operations using agent technology. In the two examples, the UNIX `who` command is demonstrated from an agent perspective. The `who` command is an operation that tells the user who else is operating on the UNIX machine that they are working on. While this may not seem like an ideal application for agents, the example is actually presenting how one might develop an agent shell. The agent that is developed in this chapter can actually perform any operation in place of the UNIX `who` command; this `who` application is just an easily understood application which has been embedded inside an Internet agent.

2.4.3 Telescript

The Telescript system, developed by General Magic, Inc., addresses the majority of issues facing the agent in the marketplace: security, mobility, and transactions. Telescript and its associated technologies are roundly acknowledged as the bases, both philosophically and technically, for most of the other agent systems in use today, and, indeed, all other systems presented in this book.

The scenario presented in the Telescript chapter is that of an agent and place for a scene from the electronic marketplace. A shopping agent, acting for a client, travels to a warehouse place. At the warehouse, the agent checks the price of a product of interest to its client, waits if necessary for the price to fall to a client-specified level, and returns when either the price is right or a specified period of time has elapsed.

2.4.4 Aglets workbench

Aglets Workbench is a mobile agent system developed by IBM Japan that is based on the Java programming language. Aglets Workbench presents the user with a visual environment for building network-centric solutions that use mobile agents to search for, access, and manage corporate data and other information.

The two main features of Aglets Workbench are its visual environment, which allows users to quickly and easily create custom agents for the Internet, and its components, which enable the agents to access corporate databases, as well as search, travel, and communicate in a standardized and secure manner.

PART II

Three complete agent systems

This section presents three different mobile agent systems that have been developed to solve many of the problems presented in Part I. Each of the systems is presented in its entirety, with concepts, driving issues in its development, and examples for how to use the mobile agents.

These systems were designed only to be agent systems, and are therefore optimized to be used in such a way. The servers, services, and agents were all designed from top to bottom to solve agent problems. This makes these systems models for the design of agents today and the future development of complete agent solutions.

In designing only for agents, the attempting to optimize all aspects of the system to this end, these systems could be referred to as closed systems. The agents written for one of the systems will only run on the servers and with the services written for that system. In the future this may be viewed as the wrong way to make a truly revolutionary agent solution, but at the present this is not a limitation in design or implementation.

As a contrast to this approach, Part III looks at an agent system based on Java and explains its advantages and differences with the above approach.

chapter 3

Telescript

CO-AUTHORED BY JAMES WHITE

Telescript was the first system to bring network agents into the public conscience. Telescript promised to be the underpinnings of a whole new model of shopping utilizing online transactions. Since Telescript was first released, the way people use the Internet and networks has changed dramatically. Throughout these changes, Telescript has withstood scrutiny and has proven itself as a language that will be around for quite awhile.

Telescript is now the basis for General Magic's Tabriz Agent System. More information on Tabriz can be found on the CD-ROM and at General Magic's Web site.[1]

In addition to the warehouse example developed in Telescript, major features of this chapter are the scenarios which are presented. There are a number of real-world situations which agents will play a major role in; these scenarios could be implemented using any of the languages in this book. It is worth reading Section 3.2, "Using mobile agents" no matter what language you are developing agents in simply because these examples are so illuminating.

3.1 Telescript mobile agent

The concept behind agents created with Telescript technology is discussed in section 2.2 and section 2.3. You can refer to these sections for an overview of the design goals for Telescript and for some of the specific terminology used in the system.

3.1.1 Object model

This section explains how a communicating application works. It does this by implementing both an agent and a place. This section begins with a discussion of the low-level concepts and terminology that underlie the example, with particular attention to the *object model,* which governs how either an agent or a place is constructed from its component parts.

Object structure

Like the first object-oriented language, SmallTalk, every piece of information, however small, is treated as an object. An *object* has both an external interface and an internal implementation.

An object's *interface* consists of attributes and operations. An *attribute,* which is an object itself, is one of an object's externally visible characteristics. An object can *get* or *set*

---

1 http://www.genmagic.com

its own attributes and the *public,* but not the *private,* attributes of other objects. An *operation* is a task that an object *performs.* An object can *request* its own operations and the *public,* but not the *private,* operations of other objects. An operation can accept objects as *arguments* and can *return* a single object as its *result.* An operation can *throw* an exception rather than return. The *exception,* an object, can be *caught* at a higher level of the agent's or place's procedure, to which control is thereby transferred.

An object's *implementation* consists of properties and methods. A *property,* an object itself, is one of an object's internal characteristics. Collectively, an object's properties constitute its dynamic state. An object can directly get or set its own properties, but not those of other objects. A *method* is a procedure that performs an operation or that gets or sets an attribute. A method can have *variables,* objects that constitute the dynamic state of the method.

Object classification

Like many object-oriented programming languages, one focus of the Telescript language is on classes. A *class* is a *slice* of an object's interface combined with a related slice of its implementation. An object is an *instance* of a class.

The programmer defines his or her communicating application as a collection of classes. To support such `user-defined` classes, the language provides many `predefined` classes, a variety of which are used by every application. The example application presented in this section consists of several user-defined classes that use various predefined classes.[2]

Classes form a hierarchy[3] whose root is Object, a predefined class. Classes other than the Object class inherit the interface and implementation slices of their superclasses. The *superclasses*[4] of a class are the root and the classes that stand between the class and the root. A class is a *subclass* of each of its superclasses. An object is a *member* of its class and each of its superclasses.

A class can both define an operation or attribute and provide a method for it. A subclass can provide an overriding method unless the class *seals* the operation. The overriding method can invoke the overridden method by *escalating* the operation using the language's "^" construct. The overriding method selects and supplies arguments to the overridden method.

---

2 The example uses the predefined classes: Agent, Class, Class Name, Dictionary, Event Process, Exception, Integer, Meeting Place, Nil, Object, Part Event, Permit, Petition, Place, Resource, String, Teleaddress, Telename, Ticket, and Time and various subclasses of Exception, such as Invalid.

3 The language permits a limited form of multiple inheritance by allowing other classes that extend the hierarchy to a directed graph.

4 The language permits a class to have implementation superclasses that differ from its interface superclasses. Such classes are rare in practice.

One operation, which the Object class defines, is subject to a special escalation rule. The `initialize` operation is requested of each new object. Each method for the operation initializes the properties of the object defined by the class that provides the method. Each method for this operation must escalate it so that all methods are invoked and all properties are initialized.

Object manipulation

Methods are required to have *references* to the objects that are to be manipulated. References serve the purpose of pointers in languages like C, but avoid the dangling pointer problem shared by such languages. References can be replicated, so there can be several references to an object.

A method receives references to the objects it creates, the arguments of the operation it implements, and the results of the operations it requests. It can also obtain references to the properties of the object it manipulates.

With a reference to an object in hand, a method can get one of the object's attributes or request one of the object's operations. It accomplishes these simple tasks with two frequently used language constructs, such as the following:

```
file.length
file.add("isEmployed", true)
```

The example application makes use of the predefined Dictionary class. A dictionary holds pairs of objects, its *keys* and *values*. Assuming that `file` denotes a dictionary, the first program fragment, `file.length`, obtains the number of key-value pairs in that dictionary, while the second program fragment adds a new pair to it. If these were fragments of a method provided by the Dictionary class itself, `file` would be replaced by `"*"`, which denotes the object being manipulated.

References are of two kinds, *protected* and *unprotected*. A method cannot modify an object to which it has only a protected reference. The engine intervenes by throwing a member of the predefined Reference Protected class.

3.1.2 Programming a place

The agent and place of the example enable this scene from the electronic marketplace. A shopping agent, acting for a client, travels to a warehouse place, checks the price of a product of interest to its client, waits if necessary for the price to fall to a client-specified level, and returns when either the price reaches that level or a client-specified period of time has

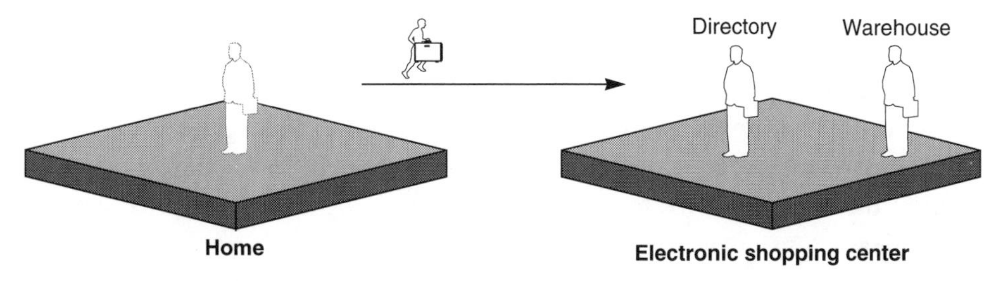

Figure 3.1 An agent goes to a place

elapsed (see Figure 3.1). The construction of the warehouse place and the client's construction and eventual debriefing of the shopping agent are beyond the scope of the example.

The warehouse place and its artifacts are implemented by three user-defined classes, discussed in the following pages.

The catalog entry class

The user-defined `CatalogEntry` class implements each entry of the warehouse's catalog, which lists the products the warehouse place offers for sale. Implicitly, in this example, this class is a subclass of the predefined Object class.

A catalog entry has two public attributes and two public operations. The `product` attribute is the name of the product the catalog entry describes, the `price` attribute is its price. The two operations are discussed after the following example.

```
CatalogEntry: class =
(
   public
      product: String;
      price: Integer; //cents
      see initialize
      see adjustPrice
   property
      lock: Resource;
);
```

The special `initialize` operation initializes the three properties of a new catalog entry. The `product` and `price` properties, implicitly set to the operation's arguments, serve as the `product` and `price` attributes. The `lock` property, set by the method to a new resource, is discussed below.

```
initialize: op (
   product: String;
   price: Integer /* cents */ ) =
{
   ^();
```

```
        lock = Resource()
};
```

A catalog entry uses a resource to serialize price modifications made using its `adjustPrice` operation. A `resource` enables what some languages call `critical conditional regions`. Here the resource is used to prevent the warehouse place and an agent of the same authority, for example, from changing a product's price simultaneously and, as a consequence, incorrectly.

The public `adjustPrice` operation adjusts the product's price by the percentage supplied as the operation's argument. A positive percentage represents a price increase, a negative percentage a price decrease.

```
adjustPrice: op (percentage: Integer)
throws ReferenceProtected =
{
   use lock
   {
      price = price + (price*percentage).quotient(100)
   }
};
```

A catalog entry, as mentioned earlier, uses a resource to serialize price modifications. Here the language's `use` construct excludes one agent or place from the block of instructions in braces, as long as another is executing them.

The operation may throw an exception. If the catalog entry is accessed using a protected reference, the engine throws a member of the predefined Reference Protected class. For example, if the shopping agent rather than the warehouse place tried to change the price, this would be the consequence.

The warehouse class

The user-defined `Warehouse` class implements the warehouse place itself. This class is a subclass of the predefined Place and Event Process classes.

A warehouse has three public operations, as shown in the following example.

```
Warehouse: class (Place, EventProcess) =
(
   public
      see initialize
      see live
      see getCatalog
   property
      catalog: Dictionary[String, CatalogEntry];
);
```

The special `initialize` operation initializes the one property of a new warehouse place. The `catalog` property, implicitly set to the operation's argument, is the warehouse place's catalog. Each key of this dictionary is assumed to equal the `product` attribute of the associated catalog entry.

```
initialize: op (
   catalog: owned Dictionary[String, CatalogEntry]) =
{
   ^()
};
```

A region can prevent a place from being constructed in that region the same way it prevents an agent from traveling there (see "Permits" in section 2.2.7). Thus a region can either prevent or allow warehouse places and can control their number.

The special `live` operation operates the warehouse place on an ongoing basis. The operation is special because the engine itself requests it of each new place. The operation gives the place autonomy. The place *sponsors* the operation—that is, performs it under its authority and subject to its permit. The operation never finishes; if it did, the engine would terminate the place.

```
live: sponsored op (cause: Exception|Nil) =
{
   loop {
      // await the first day of the month
      time: = Time();
      calendarTime: = time.asCalendarTime();
      calendarTime.month = calendarTime.month + 1;
      calendarTime.day = 1;
      *.wait(calendarTime.asTime().interval(time));

      // reduce all prices by 5%
      for product: String in catalog
      {
         try { catalog[product].adjustPrice(-5) }
         catch KeyInvalid { }
      };

      // make known the price reductions
      *.signalEvent(PriceReduction(), 'occupants)
   }
};
```

On the first of each month, unbeknownst to its customers, the warehouse place reduces by 5 percent the price of each product in its catalog. It signals this event to any agents present at the time. An `event` is an object with which one agent or place reports an incident or condition to another.

The public getCatalog operation gets the warehouse's catalog—that is, returns a reference to it. If the agent requesting the operation has the authority of the warehouse place itself, the reference is an unprotected reference. If the shopping agent requests the operation, however, the reference is protected.

```
getCatalog: op () Dictionary[String, CatalogEntry] =
{
   if sponsor.name.authority == *.name.authority {catalog}
   else {catalog.protect()@}
};
```

As mentioned earlier, one agent or place can discern the authority of another. Using the language's sponsor construct, the warehouse place obtains a reference to the agent under whose authority the catalog is requested. The place decides whether to return to the agent a protected or unprotected reference to the catalog by comparing their name attributes.

The price reduction class

The user-defined Price Reduction class implements each event that the warehouse place might signal to notify its occupants of a reduction in a product's price. This class is a subclass of the predefined Event class.

```
PriceReduction: class (Event) = ();
```

3.1.3 Programming an agent

Once it opens its doors, the warehouse needs customers. Shopping agents are implemented by the two user-defined classes presented and discussed in the following pages.

The shopper class

The user-defined Shopper class implements any number of shopping agents. This class is a subclass of the predefined Agent and Event Process classes.

A shopping agent has four public operations and two private ones, all of which are discussed individually in the following pages.

```
Shopper: class (Agent, EventProcess) =
(
   public
      see initialize
      see live
      see meeting
      see getReport
   private
```

```
    see goShopping
    see goHome
  property
    clientName: Telename; //assigned
    desiredProduct: String;
    desiredPrice, actualPrice: Integer; //cents
    exception: Exception|Nil;
);
```

The special `initialize` operation initializes the five properties of a new shopping agent. The `clientName` property, set by the operation's method to the telename of the agent creating the shopping agent, identifies its client. The `desiredProduct` and `desiredPrice` properties, implicitly set to the operation's arguments, are the name of the desired product and its desired price. The `actualPrice` property is not set initially. If the shopping agent finds the desired product at an acceptable price, it sets this property to that price. The `exception` property is set by the method to `nil`. If it fails in its mission, the agent sets this property to the exception it encountered.

```
initialize: op (
    desiredProduct: owned String;
    desiredPrice: Integer) =
{
    ^();
    clientName = sponsor.name.copy()
};
```

A region can prevent an agent from being constructed in that region the same way it prevents one from traveling there (see "Permits," earlier in the paper). Thus a region can either prevent or allow shopping agents and can control their number.

The special `live` operation operates the shopping agent on an ongoing basis. The engine requests the operation of each new agent. The new agent, like a new place, sponsors the operation and gains autonomy by virtue of it. When the agent finishes performing the operation, the engine terminates it.

```
live: sponsored op (cause: Exception|Nil) =
{
    // take note of home
    homeName: = here.name;
    homeAddress: = here.address;

    // arrange to get home
    permit: = Permit(
       (if *.permit.age     == nil {nil}
        else {(*.permit.age    *90).quotient(100)}),
       (if *.permit.charges == nil {nil}
        else {(*.permit.charges*90).quotient(100)})
    );
```

```
      // go shopping
      restrict permit
      {
        try { *.goShopping(Warehouse.name) }
        catch e: Exception { exception = e }
      }
      catch e: PermitViolated { exception = e };

      // go home
      try { *.goHome(homeName, homeAddress) }
      catch Exception { }
};
```

The shopping agent goes to the warehouse and later returns. The private `goShopping` and `goHome` operations make the two legs of the trip after the present operation records as variables the telename and teleaddress of the starting place. A *teleaddress* is data that denotes a place's network location.

Using the `restrict` construct, the shopping agent limits itself to 90 percent of its allotted time and computation. It holds the remaining 10 percent in reserve so it can get back even if the trip takes more time or energy than it had anticipated. The agent catches and records exceptions, including the one that would indicate that it had exceeded its self-imposed permit.

The private `goShopping` operation is requested by the shopping agent itself. The operation takes the agent to the warehouse place, checks the price of the requested product, waits if necessary for the price to fall to the requested level, and returns either when that level is reached or after the specified time interval. If the actual price is acceptable to its client, the agent records it.

```
goShopping: op (warehouse: ClassName)
throws ProductUnavailable =
{
   // go to the warehouse
   *.go(Ticket(nil, nil, warehouse));

   // show an interest in prices
   *.enableEvents(PriceReduction(*.name));
   *.signalEvent(PriceReduction(), 'responder);
   *.enableEvents(PriceReduction(here.name));

   // wait for the desired price
   actualPrice = desiredPrice+1;
   while actualPrice > desiredPrice
   {
     *.getEvent(nil, PriceReduction());
     try
     {
```

```
            actualPrice =
            here@Warehouse.getCatalog()[desiredProduct].price
        }
        catch KeyInvalid { throw ProductUnavailable() }
    }
};
```

The shopping agent travels to the warehouse place using the `go` operation. Upon arrival the agent expresses interest in the price reduction event that it knows the place will signal. Each time it sees a price reduction, the agent checks the product's price to see whether it was reduced sufficiently. The agent contrives one such event to prompt an initial price check. If a price reduction is insufficient, the agent waits for another.

The agent provides the `go` operation with a ticket specifying the warehouse place's class but neither its telename nor its teleaddress. In an electronic marketplace of even moderate size, this would not suffice. The agent would have to travel to a directory place to get the place's name or address, or both.

The operation may throw an exception. If the warehouse doesn't carry the product, a member of the user-defined `ProductUnavailable` class is thrown.

The private `goHome` operation is requested by the shopping agent itself. The operation returns the agent to its starting place and initiates a meeting with its client. Before initiating the meeting, the agent asks to be signaled when the meeting ends. After initiating the meeting, the agent just waits for it to end. During the meeting, the client is expected to request the `getReport` operation.

```
goHome: op (homeName: Telename; homeAddress: Teleaddress) =
{
    // drop excess baggage
    *.disableEvents();
    *.clearEvents();

    // go home
    *.go(Ticket(homeName, homeAddress));

    // meet the client
    *.enableEvents(PartEvent(clientName));
    here@MeetingPlace.meet(Petition(clientName));

    // wait for the client to end the meeting
    *.getEvent(nil, PartEvent(clientName))
};
```

The shopping agent leaves the warehouse place using the `go` operation. Before leaving it retracts its interest in price reductions and, to lighten its load, discards any notices of price reductions it received but did not examine.

The agent provides the go operation with a ticket giving the telename and teleaddress of the agent's starting place, information it recorded previously.

The special meeting operation guards the agent's report by declining all requests to meet with the shopping agent. The agent itself initiates the one meeting in which it will participate. The operation is special because the engine itself requests it whenever a meeting is requested of an agent.

```
meeting: sponsored op (
   agent: protected Telename; //assigned
   _class: protected ClassName;
   petition: protected Petition) Object|Nil
throws MeetingDenied =
{
   throw MeetingDenied();
   nil
};
```

The operation may throw an exception. Indeed the shopping agent always throws a member of the predefined MeetingDenied class.

The public getReport operation returns the actual price of the desired product. The actual price is less than or equal to the desired price.

```
getReport: op () Integer //cents
throws Exception, FeatureUnavailable =
{
   if sponsor.name != clientName
     { throw FeatureUnavailable() };
   if exception != nil
     { throw exception };
   actualPrice
};
```

The operation may throw an exception. If the agent requesting the operation is not the shopping agent's client, the operation's method throws a member of the predefined FeatureUnavailable class. If the shopping agent failed in its mission, the method throws a member of the predefined Exception class.

The product unavailable class

The user-defined ProductUnavailable class implements each exception with which the shopping agent might notify its client that the warehouse doesn't carry the product. This class is a subclass of the predefined Exception class.

```
ProductUnavailable: class (Exception) = ();
```

3.2 *Using mobile agents*

The previous section explained how communicating applications work. This final section speculates about the myriad applications that third-party developers could call into being. Each of the three subsections adopts a theme, develops one variation on that theme, and sketches four others. These are the promised scenes from the electronic marketplace of the future.

3.2.1 *Monitoring changing conditions*

These examples exploit the agents in monitoring online information that could change. Having the agent check allows the user to have the most up-to-date information while saving time and energy. Since the agent is vigilant, the user is sure to know when the information changes.

The user experience

Two weeks from now, Chris must make a two-day business trip to Boston. He makes his airline reservations using his personal communicator. Once he arrived in Boston, his schedule proves hectic. On the second day, he's running late. Two hours before his return flight is scheduled to leave, Chris's personal communicator informs him that the flight has been delayed an hour. That extra hour means that Chris doesn't have to cut short his last appointment.

Chris could have called the airline to find out whether his flight was on time, but he was extremely busy, so the hour saved was important.

How agents provide the experience

In this scenario, Chris can thank one mobile agent for booking his round-trip flight to Boston and another for monitoring his return flight and notifying him of its delay. The first of these two imaginary tasks was accomplished in the following steps (see figure 3.2).

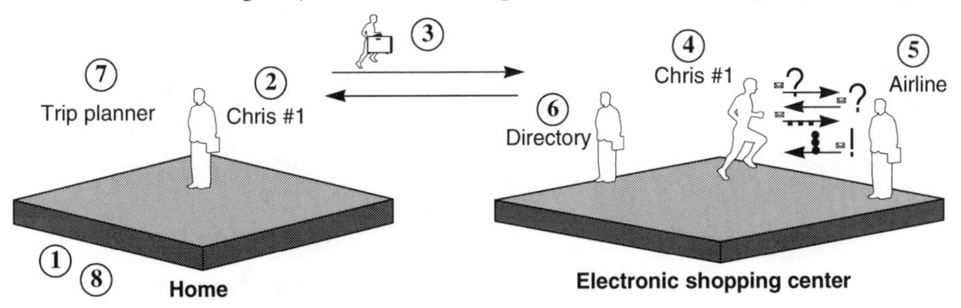

Figure 3.2 How agents work

1. Chris gives to the trip-planning application he bought for his personal communicator the dates of his trip, his means of payment (for example, his credit card or online account information), his choice of airline, and other pertinent information. If he has used the application before, it has much of this information already.
2. The application creates an agent of Chris's authority and gives his flight information to it. The part of the application written in C creates and interacts with the part written in the Telescript language, the agent, through the engine in Chris's personal communicator.
3. The agent travels from Chris's communicator to the airline place in the electronic marketplace. It does this using the `go` instruction and a ticket that designates the airline place by its authority and class.
4. The agent meets with the airline agent that resides in and provides the service of the airline place. It does this using the `meet` instruction and a petition that designates the airline agent by its authority and class.
5. The agent gives Chris's flight information to the airline agent, which compares the authority of Chris's agent to the name on Chris's online payment information and then books his flight, returning a confirmation number and itinerary.
6. The agent returns to its place in Chris's communicator. It does this using the `go` instruction and a ticket that designates that place by its telename and teleaddress, which the agent noted before leaving there.
7. The agent gives the confirmation number and itinerary to the trip-planning application. Its work complete, the agent terminates.
8. The application conveys to Chris the confirmation number and itinerary, perhaps making an entry in his electronic calendar as well.

The remaining task of monitoring Chris's return flight and informing him if it is delayed is carried out in the following additional steps (see figure 3.3).

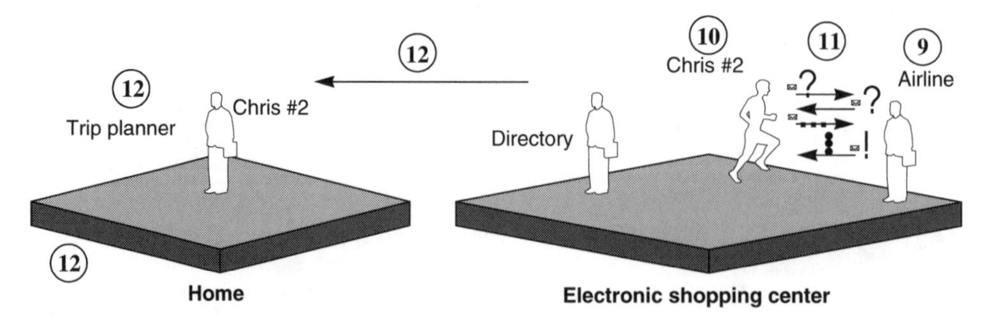

Figure 3.3 Monitoring the return

9. Before leaving the airline place (in step 6), Chris's agent creates a second agent of Chris's authority and gives his itinerary to it.
10. This second agent puts itself to sleep until the day of Chris's trip. The airline place may charge Chris a fee for the agent's room and board.
11. On the day of Chris's flight, the agent arises and checks the flight once an hour throughout the day. On each occasion it meets with the airline agent using the `meet` instruction and a petition that designates the agent by its authority and class. On one occasion it notes a delay in Chris's flight.
12. The agent returns to Chris's personal communicator (as in step 6), notifies the trip-planning application of the delay in the return flight, and then terminates (as in step 7). The application gives Chris the information that allows him to complete his meeting (as in step 8).

3.2.2 Variations on the theme

This first scenario demonstrates how mobile agents can monitor changing conditions in the electronic marketplace. There are many possible variations.

- Chris learns by chance that his favorite band will be in town next month. He tries to get tickets but learns that the concert sold out in an hour. Thereafter Chris's agent monitors the ticket exchange in the newspaper's online classifieds every morning at 6 A.M. The next time a concert is listed in his area, the agent snaps up two tickets. If Chris can't go himself, he'll sell the tickets to a friend.
- Chris buys a television from an electronics store. Chris's agent monitors the local consumer electronics market for thirty days after the purchase. If it finds the same set for sale at a lower price, the agent notifies Chris so that he can exercise the low-price guarantee of the store he patronized.
- Chris invests in several publicly traded companies. Chris's agent monitors his portfolio, sending him biweekly reports and word of any sudden stock price change. The agent also monitors the wire services, sending Chris news stories about the companies whose stock he owns.
- Mortgage rates continue to fall. Chris refinances his house at a more favorable rate. Thereafter Chris's agent monitors the local mortgage market and notifies him if rates drop 1 percent below his new rate. With banks foregoing closing costs, such a drop is Chris's signal to refinance again.

3.2.3 *Doing time-consuming legwork*

In these examples the agent performs actions for the user that do not require the user's direct interaction. There are a number of places on the Net that require a user to actively input information which could be accessed by an agent instead. The user tells the agent the information to look for and the agent decides where to look and how to gather this information.

The user experience

John is in the market for a camera. He's read the equipment reviews in the photography magazines and in consumer magazines and has visited his local camera store. He has decided on the camera he wants to buy. But from whom? John asks his personal communicator. In fifteen minutes he has the names, addresses, and phone numbers of the three shops in his area with the lowest prices. A camera store fifteen miles away offers the camera he wants at $70 below the price at his local camera shop.

Needless to say, the $70 that John saved was significant to him. He could have consulted the three telephone directories covering his vicinity, made a list of the twenty-five camera retailers within, say, twenty miles of his office, and called each to obtain its price for the camera, but who has that kind of time? John now considers his personal communicator to be an indispensable shopping tool.

How agents provide the experience

John can thank a mobile agent for finding the camera store (see figure 3.4), a task that was accomplished in the following steps

1 John gives to the shopping application he bought for his personal communicator the make and model of the camera he's selected. He also identifies the geographical area for which he wants pricing information.

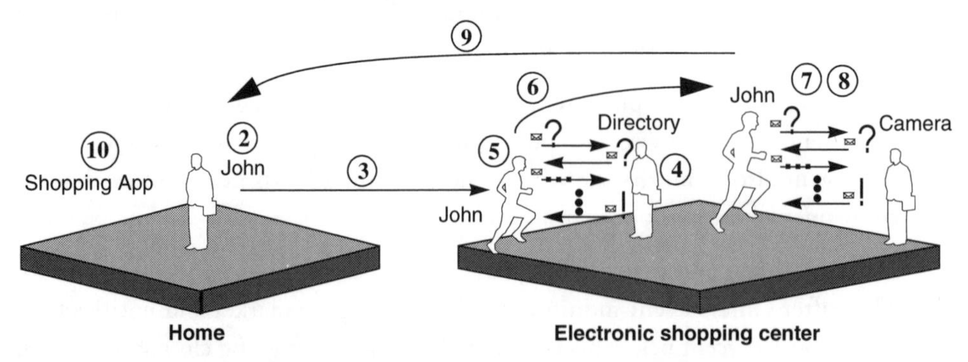

Figure 3.4 A mobile agent shops

2. The application creates an agent of John's authority and gives it John's shopping instructions. The C part of the application creates and interacts with the agent through the engine in John's communicator.
3. The agent travels from John's communicator to the directory place in the electronic marketplace. It uses the `go` instruction and a ticket that designates the directory place by its authority and class.
4. The agent meets with the directory agent that resides in and provides the service of the directory place. It uses the `meet` instruction and a petition that designates the directory agent by its authority and class.
5. The agent obtains from the directory agent the directory entries for all camera retailers about which the place has information. John's agent narrows the list to the retailers in the geographical area it is to explore.
6. The agent visits the electronic storefront of each retailer in turn. Each storefront is another place in the electronic marketplace. For each trip the agent uses the `go` instruction with a ticket that gives the `Telename` and `Teleaddress` that the agent found in the storefront's directory entry.
7. The agent meets with the camera agent it finds in each camera place it visits. It uses the `meet` instruction and a petition that designates the camera agent by its authority and class.
8. The agent gives to the camera agent the camera's make and model and is quoted a price. The agent retains information about this particular shop only if it proves a candidate for the agent's top-three list.
9. The agent eventually returns to its place in John's communicator. It does this using the `go` instruction and a ticket that designates that place by its `Telename` and `Teleaddress`, which the agent noted before leaving there.
10. The agent makes its report to the shopping application. Its work complete, the agent terminates.
11. The application presents the report to John, perhaps making an entry in his electronic diary as a permanent record.

Variations on the theme

This second scenario demonstrates how mobile agents can find and analyze information in the electronic marketplace. There are many variations.

- John hasn't talked to his college friend, Doug, in twenty years. He remembers that Doug was a computer science major. John's agent searches the trade journals and conference proceedings—even very specialized ones—in the hope that Doug has written or spoken publicly. The agent finds that Doug has published several papers, one just two years ago. The agent returns with Doug's address in Los Angeles, where he has lived for five years.

- It's Friday. John has been out of town all week on business. Expecting to go home today, John is asked to attend a Monday morning meeting in New York. He faces an unplanned weekend in midtown Manhattan. John's agent learns that his favorite tenor is performing at Radio City Music Hall on Saturday night. With John's approval, the agent purchases him a ticket for the concert.
- John is in the market for a used car. He really wants a station wagon to accommodate his growing family. John's agent checks the classified sections of all fifteen Bay Area newspapers and produces for John a tabular report that includes all used station wagons on the market. The report lists the cars by make, model, year, and mileage so that John can compare them easily.
- John yearns for a week in Hawaii. His agent voices his yearning in the electronic marketplace, giving details that John has provided: a few days on Kauai, a few more on Maui, beach-front accommodations, peace and quiet. The agent returns with a dozen packages. Unlike the junk mail that John receives by post, many of these offers are designed specifically for him. The marketplace is competing for his business.

3.2.4 Using services in combination

The following examples show agents acting in a more intelligent manner by interacting with a number of services. Although these examples may seem to exist far in the future, the behavior that is presented can be accomplished by using a number of simplistic agents. Building agents to achieve the goals presented here is possible using the agent systems in this book.

The user experience

Mary and Paul have been seeing each other for years. Both lead busy lives, and they don't have enough time together. But Mary has seen to it that they're more likely than not to spend Friday evenings together. Using her personal communicator, she's arranged that a romantic comedy is selected and ready for viewing on her television each Friday at 7 P.M., that pizza for two is delivered to her door at the same time, and that she and Paul are reminded earlier in the day of their evening together and of the movie to be screened.

Paul and Mary recognize the need to live well-rounded lives, but their demanding jobs make it difficult. Their personal communicators help them achieve their personal as well as their professional objectives. And it's fun.

How agents provide the experience

Mary relies on a mobile agent to orchestrate her Friday evenings. Born months ago, the agent waits in a quiet corner of the electronic marketplace for most of the week; each Friday at noon it springs into action (see figure 3.5).

1. Mary's agent keeps a record of the films it selected on past occasions to prevent selecting one of those films again.
2. The agent travels from its place of repose to one of the many video places in the electronic marketplace. It uses the `go` instruction and a ticket that designates the video place by its authority and class.
3. The agent meets with the video agent that resides in and provides the service of the video place. It uses the `meet` instruction and a petition that designates the video agent by its authority and class.
4. The agent asks the video agent for the catalog listing for each romantic comedy in its inventory. The agent selects a film at random from among the recent comedies, avoiding the films it has selected before. The agent orders the selected film from the video agent, charges it to Mary's account, and instructs the video agent to transmit the film to her home at 7 P.M. The video agent compares the authority of Mary's agent to the name on the account.
5. The agent goes next to the pizza place. It uses the `go` instruction and a ticket that designates the pizza place by its authority and class.

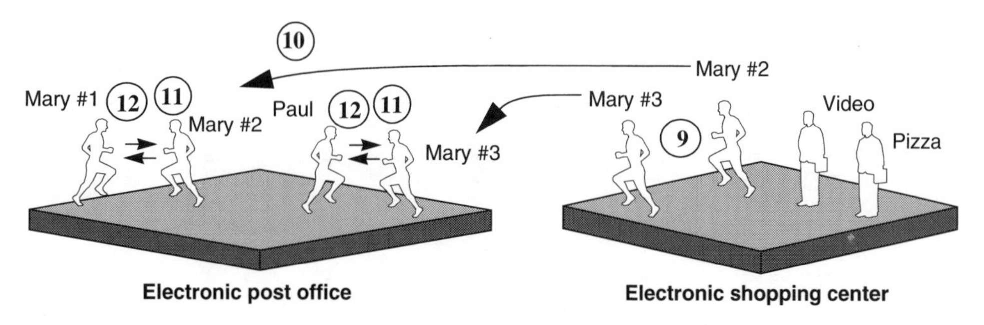

Figure 3.5 An agent plans Friday entertainment

6 The agent meets with the pizza agent that resides in and provides the service of the pizza place. It uses the `meet` instruction and a petition that designates the pizza agent by its authority and class.

7 The agent orders one medium-size pepperoni pizza for home delivery at 6:45 P.M. The agent charges the pizza, as it did the video, to Mary's online account. The pizza agent, like the video agent before it, compares the authority of Mary's agent to the name on the agent's account.

8 Mary's agent returns to its designated resting place in the electronic marketplace. It uses the `go` instruction and a ticket that designates that place by its `Telename` and `Teleaddress`, which it noted previously.

All that remains is for the agent to notify Mary and Paul of their evening appointment (see figure 3.6). This is accomplished in the following additional steps.

9 The agent creates two new agents of Mary's authority and gives each the catalog listing of the selected film and Mary's and Paul's names. Its work complete, the original agent awaits another Friday.

10 One of the two new agents goes to Mary's mailbox place and the other goes to Paul's. To do this they use the `go` instruction and tickets that designate the mailbox places by their class and authorities.

11 The agents meet with the mailbox agents that reside in and provide the services of the mailbox places. They use the `meet` instruction and petitions designating the mailbox agents by their class and authorities.

12 The agents deliver to the mailbox agents electronic messages that include the film's catalog listing and that remind Mary and Paul of their date. The two agents terminate and the mailbox agents convey the reminders to Mary and Paul.

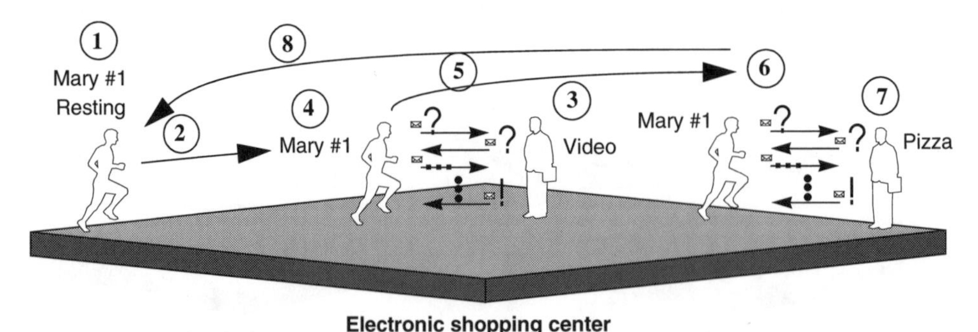

Figure 3.6 Notifying the participants

Variations on the theme

This third scenario demonstrates how mobile agents can combine existing services to create new, more specialized services. There are many variations.

- Mary plans to take Paul to see a stage show next weekend. Her agent tries to book orchestra seats for either Saturday or Sunday, gets them for Sunday, reserves a table at a highly regarded Indian restaurant within walking distance of the theater, and orders a dozen roses for delivery to Paul's apartment that Sunday morning.
- Mary's travel plans change unexpectedly. Rather than returning home this evening as planned, she's off to Denver. Mary's agent alters her airline reservation, books her a nonsmoking room at a hotel within fifteen minutes of her meeting, reserves her a compact car, and provides her with driving instructions to the hotel and the meeting. The agent also supplies Mary with a list of Indian restaurants in the vicinity.
- Mary receives and pays her bills electronically. Her agent receives each bill as it arrives, verifies that Mary has authorized its payment, checks that it is in the expected range, and issues instructions to the bank. Mary's agent prepares for her a consolidated monthly report, and at tax time sends her accountant a report of her deductible expenses.
- Mary subscribes to a daily newspaper, but it isn't her idea of news. At 7 A.M. every day Mary's agent delivers to her a personalized newspaper. It includes synopses of the major national and international news stories of course, but it also reports the local news from her hometown in Virginia; the major events of yesterday in her field, physics; and the market activity of the stocks in her portfolio.

chapter 4

Agent Tcl

CO-AUTHORED BY ROBERT S. GRAY

The Agent Tcl system has four main features. It makes it easy for agents to move around by reducing migration to a single instruction, `agent_jump`. This instruction can be used at any point by the agent during its operation; the agent can jump whenever it wants. It provides a way for communication that is flexible and low-level. It provides a high-level scripting language as the main agent language. And it provides an effective security model, which is crucial to the agents' acceptance in the agent world.

In addition to the well-developed architecture which provides the basis of the Agent Tcl system, another of Agent Tcl's real powers comes from its usage of the Tcl scripting language. Tcl is a widely used, easy to learn scripting language which presents its operations at a high-level for the user. By basing agents on the Tcl language, users are presented with easier methods for learning and deploying powerful agents.

The examples that are presented for Agent Tcl show an agent that travels around a network looking for who is working on each machine; this information is sent back to the agent's home. The agent that is developed can be modified very easily to solve any number of network agent problems.

4.1 Overview

Agent Tcl is a powerful Internet agent system that runs on UNIX workstations and allows the rapid development of complex agents. Agent Tcl is an effective platform for experimentation with Internet agents and for the development of small- to medium-sized applications, although it may lack some of the more advanced features of a commercial agent system such as General Magic's Telescript. Agent Tcl agents are written in an extended version of the Tool Command Language (Tcl) developed by Sun Microsystems.[1] Tcl is a high-level scripting language that is both powerful and easy to learn. It was designed initially to allow programmers to tie together various applications and utilities on UNIX machines in order to provide more comprehensive solutions for users. This makes Tcl an ideal language for Internet agents because most are concerned primarily with coordinating high-level communication and resource access. Agent Tcl agents can use all of the standard Tcl commands as well as a set of special commands that are provided as a Tcl extension. These special commands allow an agent to migrate from one machine to another, to create child agents, to communicate with other agents, and to obtain information about the agent's current network location. In addition, Agent Tcl, like all Tcl-based systems, can be extended with user-defined commands to create a more powerful agent system—for example, a set of text-processing commands can be

1 Information of Tcl can be found in *Tcl and the Tk Toolkit*, by John K. Ousterhout.

made available to all agents at a particular site. This provides developers with the ability to come up with services for users which can be easily accessed by the agents.

4.1.1 Agent movement

The real power of Internet agents comes from the ability to move from machine to machine. In Agent Tcl, this migration is accomplished with the `agent_jump` command, which can appear anywhere within an agent. It captures the current state of the agent and transfers this state image to a server on the destination machine. The server restores the state image and the agent continues execution from the command immediately after the `agent_jump`. In other words, `agent_jump` allows the agent to suspend its execution at an arbitrary point, transport to another machine, and resume execution on the new machine at the exact point at which it left off. This approach to migration is the same as in Telescript, considered by many to be the best-developed Internet agent language at this time. Once an Agent Tcl agent has migrated to a machine, it can access resources and communicate with other agents on that machine. Once it finishes its local task, it can migrate to the next machine.

4.1.2 Agent communication

Once an agent is at a particular place, it goes about its tasks by communicating with others at the site. There are two ways for agents to communicate with each other. The first is message passing, which uses the traditional *send* and *receive* concepts. The `agent_send` command sends a message to another agent and the `agent_receive` command receives an incoming message. The second form of communication is a direct connection, which is essentially a named message stream. An agent establishes a direct connection with another agent using the `agent_meet` command. The two agents then exchange messages over the connection. Direct connections are more efficient than message passing for long interactions and are convenient for the programmer since the agent can wait for messages on a particular connection. A message in Agent Tcl is an arbitrary string with no predefined syntax or semantics, so the agents must agree on the meaning of the messages that they exchange. The base communication mechanisms were made purposely low-level to allow experimentation with a range of communication paradigms. Two paradigms have already been implemented on top of the base facilities. The first is analogous to RPC (Remote Procedure Call); the second is a conversational approach that views communication between a pair of agents as an ongoing dialog. An agent can participate in as many simultaneous dialogs as desired. Each dialog has its own state space; incoming messages from other agents are automatically turned into events that execute within the appropriate state space.

4.1.3 Other agent Tcl features

Other commands in Agent Tcl allow an agent to create child agents and to obtain information about the current machine, such as the identities of other agents. In addition, agents can use the Tk toolkit to interact with the user of the current machine. Tk is a tool for creating graphical user interfaces (GUIs) for Tcl programs. Tk supports all of the standard GUI features such as windows, menus, scrollbars, and drawing areas. Since Tk allows a GUI to be written entirely in Tcl, professional-quality interfaces can be created with a relatively small amount of time and code. An agent that wanted to interact with a user during the course of its travels would carry the necessary Tk code with it. Once it reached the user's machine, it would execute the code and present the interface.

4.1.4 State of agent Tcl

Agent Tcl is an ongoing research project at Dartmouth College and is in continuous development. Research at Dartmouth focuses on the security issues associated with roving code and on support for mobile computing, since agents become particularly useful when they can migrate to and from portable machines. Portable machines are often disconnected from their network and often have an unreliable, low-bandwidth connection when they are connected. By migrating, an agent can avoid extensive use of the poor connection. These goals will be seen throughout this chapter as driving factors for many of the design decisions that have been made for the system.

This chapter describes the present state of the Agent Tcl system, which is the version that is included on the CD-ROM. The chapter also presents the future of the architecture as planned. Potential uses for Agent Tcl are presented with a specific programming example in which an agent collects system information from each machine that it visits. It concludes with a discussion of the weaknesses and strengths of Agent Tcl and its future outlook.

4.2 Architecture

Agent Tcl has four main goals:

- Reduce migration to a single instruction, `agent_jump`, and allow this instruction to occur at arbitrary points. The instruction should capture the complete state of the agent and transparently send this state to the destination machine.

The programmer should not have to explicitly collect state information, and the system should hide all transmission details even if the destination machine is a mobile computer that is temporarily disconnected or has a new network address.

- Provide communication mechanisms that are flexible and low-level, but that hide all transmission details, including whether the agents are on the same or different machines.
- Provide a high-level scripting language as the main agent language, but support multiple languages and transport mechanisms, and allow the straightforward addition of a new language or transport mechanism. Multiple languages are particularly important since, although a high-level scripting language such as Tcl is appropriate for most Internet agents, it is ill-suited for agents that require large amounts of code or that perform speed-critical tasks.
- Provide effective security in the uncertain world of the Internet.

The overall goal is a simple, flexible, and secure Internet agent system that will allow the programmer to select the most appropriate language for the task and rapidly develop even large-scale applications.

4.3 The agent Tcl architecture

The final architecture for Agent Tcl is shown in Figure 4.1. This architecture builds on the server model of Telescript, the multiple languages of Ara, and the transport mechanisms of two predecessor systems at Dartmouth. The architecture has four levels. The lowest level is an application programming interface (API) for the available transport mechanisms. The second level is a server that runs at each network site to which agents can be sent. The server performs the following tasks:

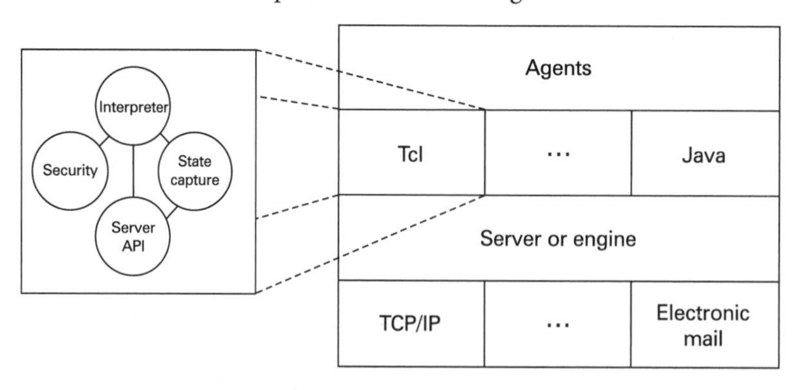

Figure 4.1 The architecture of Agent Tcl

- *Status.* The server keeps track of the agents that are running on its machine and answers queries about their status.
- *Migration.* The server accepts each incoming agent, authenticates the identity of the owner, and passes the authenticated agent to the appropriate interpreter for execution. The server selects the best transport mechanism for each outgoing agent.
- *Communication.* The server provides a hierarchical namespace for agents and allows agents to send messages to each other within this namespace. The topmost division of the namespace is the symbolic name of the agent's network location. A message is an arbitrary sequence of bytes with no predefined syntax or semantics except for two types of distinguished messages. An *event* message provides asynchronous notification of an important occurrence while a *connection* message requests or rejects the establishment of a direct connection. A direct connection is a named message stream between agents and is more convenient and efficient than message passing for long interactions (since the programmer can wait for messages on a particular stream and the server often can hand control of the stream to the interpreter). The server buffers incoming messages, selects the best transport mechanism for outgoing messages, and creates a named message stream once a connection request has been accepted.
- *Nonvolatile store.* The server provides access to a nonvolatile store so that agents can back up their internal state as desired. The server restores the agents from the nonvolatile store in the event of machine failure.

All other services are provided by agents. This approach provides the most flexibility and, with sufficient engineering work on the inter-agent communication mechanisms, should be nearly as efficient as providing the services directly in the agent servers. Such services include resource discovery, group communication, fault tolerance, access control, network sensing, and location-independent communication. The most important service agents in the internal Dartmouth prototype are *docking* agents and *resource-manager* agents. Docking agents support disconnected operation; if an agent is unable to migrate to the desired location because of machine or network failure, the agent is added to a queue or dock within the network. The dock forwards the agent to the desired location once it becomes reachable. Resource-manager agents, in combination with the Pretty Good Privacy (PGP) encryption system and language-specific security modules such as Safe-Tcl, guard access to critical system resources such as the screen, disk, and speaker. PGP authenticates incoming agents; the resource managers assign access restrictions based on this authentication; and Safe-Tcl enforces the access restrictions. In other words, the resource-manager agents provide the security policy, while Safe-Tcl provides the enforcement mechanism. This approach means that the same resource managers can provide the security policy for any agent, regardless of the agent's implementation language. Only the enforcement mechanism needs to change from one language to another.

The third level of the Agent Tcl architecture consists of one interpreter for each available language. It is called interpreter since it is expected that most of the languages will be interpreted due to portability and security constraints (although "just-in-time" compilation is feasible for languages such as Java). Each interpreter has four components—the interpreter itself, a security module that prevents the agent from taking malicious action, a state module that captures and restores the internal state of an executing agent, and an API that interacts with the server to handle migration, communication, and checkpointing. Adding a new language consists of writing the security module, the state-capture module, and a language-specific wrapper for the generic API. The security module does not determine access restrictions but instead ensures that an agent does not bypass the resource managers or violate the restrictions imposed by the resource managers; the security module for Tcl agents is the existing Safe Tcl extension that allows a Tcl interpreter to replace "dangerous" commands with safe equivalents that perform access checks. The state-capture module must provide two functions for use in the generic API. The first, `captureState`, takes an interpreter instance and constructs a machine-independent byte sequence that represents its internal state. The second, `restoreState`, takes the byte sequence and restores the internal state.

The top level of the Agent Tcl architecture consists of the agents themselves.

4.4 Current status

The architecture presented above has not been completely implemented at this time. Many of the more advanced and more esoteric features have not been completed at this time. What has been implemented is the full set of features required to develop and deploy robust and powerful Internet agents. The current implementation does not provide the nonvolatile store or multiple languages and transport mechanisms, although the framework for incorporating additional languages and transport mechanisms is in place. In addition, several components were undergoing final revision and testing at the time of publication and were not ready for public release.

Therefore, the CD-ROM contains the "stripped-down" version of Agent Tcl that was described in the introduction. This version has the following features:

- There is a single language (Tcl) and a single transport mechanism (TCP/IP). Agents can use all of the standard Tcl features as well as the Tk toolkit.

- Migration, message passing, and direct connections are supported, although the syntax of direct connections is artificially tied to the TCP/IP protocol.
- The namespace is flat rather than hierarchical.
- The docking and resource-manager agents and the authentication subsystem are not included. This means that there is no direct support for mobile computers and that the security mechanisms are rudimentary. The security mechanisms are sufficient, however, for experimentation and for local applications—that is, an agent server will only accept an incoming agent or message if it originated from an approved machine, a list of which is given to each server at startup.
- An agent server can only provide limited status information about the agents that are running on its machine.

By the time that this book appears on shelves, newer versions will be available for download. Interested readers should refer to the downloading instructions in the Agent Tcl section of the CD-ROM. The current version has proven to be a useful tool both at Dartmouth and at several external sites even without some of the features mentioned above. Part of its usefulness comes from the selection of Tcl as the main agent language. The rest of this section presents the rationale behind the use of Tcl and the details of how a Tcl script interacts with the agent system. The subsequent sections present existing and potential applications for Agent Tcl and a specific programming example.

4.5 Tcl

Agent Tcl relies on the Tcl scripting language for developing agents. Tcl is a high-level scripting language that was developed in 1987 at Sun Microsystems and has enjoyed enormous popularity in the UNIX community.[2] Tcl has several advantages as an Internet agent language, as well as some disadvantages.

4.5.1 Advantages

Tcl is easy to learn and use due to its elegant simplicity and an imperative style. Tcl is interpreted, so it is highly portable and easier to make secure. It can be embedded in

2 Tcl and Tk are now available for Macintosh and PC computers, although Agent Tcl does not run on these systems at this time.

other applications, which allows these applications to implement *part* of their functionality with mobile Tcl agents. Finally, Tcl can be extended with user-defined commands, which makes it easy to tightly integrate agent functionality with the rest of the language and allows a resource to provide a package of Tcl commands that an agent uses to access the resource. A package of Tcl commands is more efficient than encapsulating the resource within an agent and is an attractive alternative in certain applications.

4.5.2 Disadvantages

In terms of disadvantages, Tcl is inefficient compared to other interpreted languages and in orders of magnitude slower than optimized C. It provides no code modularization aside from procedures, which makes it difficult to write and debug large scripts. These disadvantages have not been a hindrance so far since Internet agents tend to involve high-level resource access wrapped with straightforward control logic, a situation for which Tcl is uniquely suited. An Internet Tcl agent is usually short even if it performs a complex task, and is usually more than efficient enough when compared to resource and network latencies. In addition, several groups are working on structured-programming extensions to Tcl and on faster Tcl interpreters.

Tcl is not suitable for every Internet agent application, however, such as an application that performs search operations against large, distributed collections of numerical data. For this reason Agent Tcl includes a framework for incorporating additional languages. This framework is being used to add support for the new Java language.

Java is much more structured than Tcl and has the potential to run at near-native speed through just-in-time compilation. It is expected, however, that Tcl will continue to be the main agent language and that Java will be used only for speed-critical agents (or portions of agents).

The main disadvantage of Tcl is that it provides no facilities for capturing the internal state of an executing script. Such facilities are essential for providing transparent migration at arbitrary points.

Adding these facilities to Tcl was straightforward but required the modification of the standard Tcl interpreter. The basic problem is that the Tcl interpreter evaluates a script by making *recursive* calls to a function called `Tcl_Eval`. The handler for the `while` command, for example, recursively calls `Tcl_Eval` to evaluate the body of the loop. Thus a portion of the script's state is on the interpreter's runtime stack and is not easily accessible. Agent Tcl's solution adds an explicit stack to the Tcl interpreter.

The command handlers are split into one or more subhandlers where there is one subhandler for each code section before or after a call to `Tcl_Eval`. Each call to `Tcl_Eval` is replaced with a push onto the stack. `Tcl_Eval` iterates until the stack is empty and always calls the current subhandler for the command at the top of the stack. The subhandlers are responsible for specifying when the command has finished and should be popped off the stack.

The explicit stack is different than the Agent for Remote Access (Ara) solution in which the C runtime stack must be captured in a portable way and the Tcl interpreter on the destination machine must contain the same set of C functions. On the other hand, the explicit stack is less efficient. Agent Tcl's modified Tcl core runs Tcl scripts approximately 20 percent slower than the standard Tcl interpreter, whereas Ara's modified Tcl interpreter imposes no additional overhead.

Once the explicit stack was available, it became trivial to write procedures that save and restore the internal state of a Tcl script. These two procedures, `captureState` and `restoreState`, are the heart of the state-capture module for the Tcl interpreter. They capture and restore the stack, the procedure call frames, and all defined variables and procedures. Such things as open files and linked variables are ignored.

The advantages of Tcl are strong and the disadvantages are either easily overcome or do not affect most agents. Thus Tcl was chosen as the main language for the Agent Tcl system. The same advantages have led to the use of Tcl in other Internet agent systems such as Ara.

4.6 *Tcl scripts as agents*

An Agent Tcl agent is a Tcl script that runs on top of the modified interpreter and a Tcl extension. The modified interpreter provides the explicit stack and the state-capture routines. The extension provides the set of commands that the script uses to migrate, communicate, and create child agents.

Due to the nature of Tcl extensions, these commands are tightly integrated with the normal Tcl commands, and, in fact, appear to be a part of the language itself. Internally each command uses the generic server API to contact an agent server, transfer an agent, message or request, and wait for a response.

The main difference between the current and planned implementations is that when migrating, creating a child agent, or sending a message, the current implementation bypasses the local server and interacts directly with the destination server over TCP/IP. This approach was adopted to simplify the initial implementation and will change as additional transport mechanisms are added.

The most important agent commands are:

- `agent_begin`
- `agent_submit`
- `agent_jump`

- agent_send
- agent_receive
- agent_meet
- agent_accept
- agent_end

An agent uses the agent_begin command to register with a server and obtain an identifier in the flat namespace. An identifier currently consists of the IP address of the server, a unique integer, and an optional symbolic name that the agent specifies later with the agent_name command.

The agent_submit command is used to create a child agent on a particular machine. The agent_jump command migrates an agent to a particular machine. The command captures the internal state of the agent, packages the state image for transport, and sends the state image to the destination server. The server accepts the state image, selects a new identifier for the agent, and starts a Tcl interpreter. The Tcl interpreter restores the state image and resumes agent execution at the statement immediately after the agent_jump.

The agent_send and agent_receive commands are used to send and receive messages. The agent_meet and agent_accept commands are used to establish a direct connection between agents. For direct connections, the source agent uses agent_meet to send a connection request to the destination agent. The destination agent uses agent_accept to receive the connection request and send either an acceptance or rejection.

An acceptance includes a TCP/IP port number to which the source agent connects. The protocol works even if both agents use agent_meet. The agent with the lower IP address and integer identifier selects the port to which the other agent connects. The agent server will take on more of the responsibility for establishing a direct connection as additional transport mechanisms are added.

4.7 Examples

Internet agents are best viewed as a tool for developing distributed applications. This feature, along with the development of more advanced interface agents and the addition of powerful and far-reaching networks, makes Internet agents a very enabling technology.

Just as with any new technology, the kill-application is often hard to define; if it were easy everyone would be doing it. Internet agents are no exception to this problem. What is presented here are the initial applications which have been developed using the Agent Tcl system. None may seem like the killer-app, but then none of them is supposed to be. Instead, these applications are versions of numerous existing applications which have been reimplemented using Internet agents. With the experience gained in the development of these systems, and the deployment of more Internet agent systems to support these systems, developers and users will get a better idea of where to go next in the pursuit of the ultimate application.

4.7.1 Possible applications

Some initial applications for Agent Tcl come from those that exist in other Internet agent systems. The Telescript system, for example, is currently used in active mail, network and platform management, and electronic commerce. In active mail, a program is embedded inside a mail message. This program is executed when the mail message is received or viewed. This embedded program can be a Telescript agent. In one platform-management application, a Telescript agent is used to perform software updates; the agent carries the necessary files onto the machine, installs the files and then disappears.

Owners of a Sony MagicLink or a Motorola Envoy have received several software updates this way. (The MagicLink and Envoy are two personal digital assistants that are based around Telescript and the MagicCap operating system.) In several electronic-commerce applications, a Telescript agent leaves a personal digital assistant (PDA), searches multiple electronic catalogs for a certain product, and returns to the PDA with the best purchase price and the corresponding vendor.

Java applets also suggest many potential applications. A Java applet is (usually) an interactive, graphical application that is automatically brought to and executed on a user's machine when the user visits the applet's enclosing Web page. Existing Java applets include stock tickers, games, and language tutorials. Java applets would be intolerably slow if they controlled the screen from a remote location; dynamic deployment allows them to control the screen efficiently without the need for pre-installation. Internet agents can play the same role as Java applets by carrying interface code to the user's location.

4.7.2 Present applications

Agent Tcl is well suited to most of the above applications. It would have some trouble with the network and platform-management applications of Telescript since Tcl has no direct capabilities for working with binary data.

Presently, Agent Tcl is being used in three information-retrieval applications. The first involves searching distributed collections of technical reports; the second, medical records; and the third, three-dimensional drawings of mechanical parts.

In each application, there is a collection of documents at one or more network sites. Each collection provides a set of low-level search primitives. Agents use these primitives to perform a multi-step search at each site. Since the agents move to the location of the collection and do not transfer intermediate results, the multi-step searches can be performed efficiently even though only low-level primitives are available.

In addition, since the agent does not need to be in continuous contact with the user's machine, it can continue its task even if the user's machine becomes temporarily unreachable.

Agent Tcl is also being used in several workflow applications, although these are less mature than the information-retrieval applications. In one application, an agent carries an electronic form from machine to machine so that the appropriate people can fill out their sections of the form, presented using Tk. In a second application, Agent Tcl handles purchase orders. An independent traveling salesperson carries a laptop with software that helps to select vendors and products and to place orders. Agents are sent to search vendor catalogs for products that meet customer needs. When a product and vendor is selected, an agent travels to the vendor's computers where it interacts with billing, inventory, and shipping agents to arrange the purchase.

In both cases, the agents can continue working even while the laptop is disconnected. This application is easier to implement with some support agents for mobile computing that are not included on the enclosed CD-ROM, although a simple implementation can be created without these support agents. Agent Tcl is also being used outside Dartmouth, most notably to execute complex queries against remote databases.

4.8 Agent design

In order to fully understand how Internet agents are different from presently existing programming solutions and how to use them effectively, it is important to understand the underpinnings of the agent systems. In order to do this, each of the chapters from here on explains the language and architecture of the agent system being presented. Much of this information is technical and may seem at times to be rather esoteric, not to mention boring.

4.9 Language design

Agent Tcl agents are written in the Tool Command Language (Tcl). Tcl has two components. The first is a shell, usually called `tclsh`, that is used to execute stand-alone Tcl scripts and interactive commands. The second is a library of C functions. The library provides functions to create a Tcl interpreter, define new commands in the interpreter, and submit Tcl scripts to the interpreter for evaluation. This library allows Tcl to be embedded inside a larger application; any application that needs a scripting language can include the library and allow its users to write Tcl scripts.

4.9.1 A factorial example

A tutorial on Tcl is beyond the scope of this chapter. Tcl is easy to learn, however, and is similar to other scripting languages such as Perl and the various UNIX shells. The following script, for example, asks the user for a number and then displays the factorial of that number. The script keeps asking for numbers until the user enters `Q` to stop.

For now, simply examine the key features of the script; how to actually run the script is described in the next section.

```
    # Procedure "factorial" recursively computes a factorial.

proc factorial x {
    if {$x <= 1} {
        return 1
    }
    return [expr $x * [factorial [expr $x - 1]]]
}

    # Repeat until the user enters "Q" to quit.

set number ""
while {$number != "q"} {

        # Get the integer for which we want the factorial
        # (or "Q" to quit).

    puts -nonewline
      "Enter a nonnegative integer (or "Q" to quit): "
    gets stdin number

        # Convert to lowercase in case it's a "Q".

    set number [string tolower $number]

        # Compute the factorial if we're not quitting.

    if {$number != "q"} {
        puts "$number! is equal to [factorial $number]"
    }
}
```

4.9.2 *Strings*

Tcl stores all data as strings. The `number` variable, for example, can be used to hold both a number and the letter `Q` because Tcl stores numbers as strings. Commands that expect numbers, such as `expr` (which evaluates general mathematical expressions), convert the given strings into an internal numeric representation.

4.9.3 *Grammar*

Tcl has no fixed grammar that defines the language. The interpreter does not treat the `while` construct above, for example, as a reserved word, followed by an expression, followed by a repeatedly executed subprogram. Instead the interpreter treats the construct as a command name, `while`, followed by two argument strings; the curly bracket characters, `{and}`, represent nothing more than a kind of string quotation. The two arguments are passed to the handler for the `while` command which interprets them as it sees fit.

The standard `while` handler does, in fact, treat the first argument as an expression, and if the expression is true, passes the second argument back to the Tcl interpreter for evaluation as a script. If the `while` handler is replaced, however, the behavior of the `while` command changes. Thus, although many Tcl commands look and act like traditional programming constructs, it is important to remember that Tcl parses everything as a command name and arguments.

4.9.4 *Substitutions*

There are two types of special syntactic constructs that can appear inside the argument strings. These constructs are called *substitutions*. In the command `expr $x * [factorial [expr $x - 1]]`, for example, `$x` is a variable substitution, and `[expr $x - 1]` is a command substitution. When the command is parsed, `$x` will be replaced with the *contents* of variable `x`, and `[expr $x - 1]` will be replaced with the *result* of executing the command `expr $x - 1`, namely the value of `$x - 1`.

The quotation characters around the string determine whether these substitutions are actually performed. Curly brackets, for example, mean that substitutions are not performed and that the string is passed unchanged to the command handler. Double quotes (`"`) or no quotes means that substitutions are performed.

In the preceding `while` command, curly brackets are used around the first argument, `$number != "q"`, so that the string is passed unchanged to the `while` handler.

The variable substitution $number is then performed once per iteration, each time that the while handler checks the value of the expression.

If double quotes had been used instead, the variable substitution would have been performed when the while command was first parsed, and the string passed to the while handler would have been "" != "q". This expression is always true so the loop would have run forever.

Proper quoting is the most difficult aspect of Tcl; it will be easier if you remember that the Tcl interpreter parses everything as a string, and that the different quotation characters affect the parsing process.

Keeping these three points in mind, it becomes straightforward to understand the script. First, the proc command is used to create a new command called factorial that takes a single argument x and computes x! by making recursive calls to itself. Then, the puts and gets commands are used to interact with the user and obtain a number; the factorial command is called with this number as its argument; and puts is used to display the factorial result.

The while command repeats this process until the user enters Q rather than a number.

This script highlights the main features of Tcl but uses only a small fraction of the Tcl commands. More information on Tcl can be found in the books by Ousterhout and Welch, in the man pages (a type of UNIX help system) that are included on the CD-ROM, and in the comp.lang.tcl Usenet group.

4.9.5 Agent creation commands

In addition to the standard Tcl commands, Agent Tcl agents use a special set to migrate from machine to machine and to communicate with other agents. These commands are provided as a Tcl extension, but can be treated as a native part of the Tcl language when writing an agent. In the remainder of this section, each command is briefly defined. In the next section, the commands are used to develop two agents. The commands can be divided into three main categories. The first category of commands allow an agent to register itself with an agent server and to obtain an identifier in the agent namespace.

- agent_begin [*machine*]. The agent_begin command registers the agent with the agent server on the specified machine (or on the local machine if no machine is specified) and returns the agent's new identifier within the agent namespace. In the current system, this identifier consists of the symbolic name of the server, the IP address of the server, a symbolic name that the agent chooses for itself, and a unique integer that the server assigns to the agent. So if an agent issues the command agent_begin bald, for example, the command might return the four-element Tcl list `bald.cs.dartmouth.edu 129.170.192.98 {} 15`. The `129.170.192.98` is the IP address of `bald`.

- The empty curly brackets indicate that the agent initially has no symbolic name; a symbolic name can be chosen at a later time with the `agent_name` command. The `15` is the integer ID that the server on `bald` has assigned to the new agent; this integer is unique among all agents executing on `bald` but not among all agents everywhere.

  The agent's current identifier is stored in element `local` of the global Tcl array `agent`. This array is always available inside an Agent Tcl script and is read-only; it contains other useful information as will be seen in the programming examples below. Once the agent has issued the `agent_begin` command, it can use the other agent commands.

- `agent_name` *name*. The `agent_name` command selects a symbolic name for the agent. If the agent in the example above issues the command `agent_name FtpAgent`, its complete name will become `bald.cs.dartmouth.edu 129.170.192.98 FtpAgent 15`.

- `agent_end`. An agent calls `agent_end` command when it is finished with its task and no longer requires agent services.

4.9.6 *Agent migration commands*

The second category allows an agent to migrate from machine to machine and to create child agents.

- `agent_jump` *machine*. An agent calls the `agent_jump` command when it wants to migrate to a new machine. This command captures the internal state of the agent and sends the state to the agent server on the specified machine. The server restores the state and the agent continues execution immediately after the `agent_jump`. Certain components of the state, such as open files and child processes, are intrinsically tied to a specific machine and are not transferred to the new machine. The agent receives a new 4-element identification when it jumps, which again is stored in element `local` of the global Tcl array `agent`. The agent also loses its symbolic name when it jumps and must request it again if needed.

- `agent_fork` *machine*. The `agent_fork` command is roughly analogous to UNIX `fork`. It creates a copy of the agent on the specified machine. Both the original agent and the copy continue execution from the point of the `agent_fork`. The `agent_fork` command returns the 4-element identification of the copy to the original agent and the string `CHILD` to the copy.

- `agent_submit` *machine* `-procs` *names* `-vars` *names* `-script` *script*. The `agent_submit` command creates a completely new agent. The parameters to `agent_submit` are a machine, a list of Tcl variables, a list of Tcl procedures, and a startup script. A new agent is created on the specified machine. This agent contains copies of the specified variables and procedures and begins execution by evaluating the startup script. The `agent_submit` command returns the 4-element identification of the new agent.

4.9.7 Agent communcation commands

The final category of commands allow agents to communicate with each other.

- `agent_send` *id code string*. The `agent_send` command sends a message to another agent. A message consists of an integer *code* and an arbitrary *string*. The recipient agent is specified by its 4-element *id* or by any subset of the 4-element ID that uniquely identifies the agent, such as the server name and the unique integer. The recipient receives the message using the `agent_receive` command, or if it is using Tk, by establishing an event handler for incoming messages using the `mask` command.
- `agent_event` *id tag string*. The `agent_event` command is a variant of `agent_send` that sends a *tag* and a *string* rather than an integer *code* and a *string*. A tag is just an arbitrary string itself. The advantage of `agent_event` is that the recipient can associate event handlers with specific tags using the `mask` command. The event handler is called automatically when a message arrives with the corresponding tag. If the recipient is not using Tk or chooses not to use event handlers, it must receive these tagged messages with the `agent_getevent` command.
- `agent_meet` *id*. The `agent_meet` command is used to request a direct connection with the specified recipient. The recipient accepts the connection request either by issuing its own `agent_meet` command or with the `agent_accept` command. Once the connection request has been accepted, and the direct connection has been established, arbitrary data can be sent along the connection using the `tcpip_read` and `tcpip_write` commands. The names of these commands reflect the current link between direct connections and TCP/IP; they should be changed but have been left alone for backward compatibility. Direct connections are more efficient than the two message-passing variants since they bypass the agent servers.

There are several miscellaneous commands that do not fall into the three main categories. The `agent_info` command, for example, is used to obtain information from a server about the agents executing on its machine; the `retry` command retries a block of Tcl code until no error occurs or the maximum number of tries has been reached; and the `restrict` command imposes a timeout on an arbitrary block of Tcl code. The documentation on the enclosed CD-ROM describes these commands, along with all of the commands listed above, in more detail.

4.10 Programming examples

There are two examples presented here, both based on a presently existing utility for UNIX machines. Why develop an agent for a problem that already has a solution? As mentioned earlier, this is the first step in the adoption of any new technology. Additionally, these two examples present one of the main features of Internet agent systems.

Much of the work that an Internet agent does has nothing to do with the problem being solved. As stupid as this may seem, it is one of the real keys to understanding Internet agents. The four goals for Agent Tcl presented at the beginning of the chapter make no mention of the type of problems that can be solved. Instead, they outline a way to move agents around in order to solve problems. Internet agent systems give developers a way to solve problems by having their systems move around from machine to machine, easily, in a secure manner, in order to solve problems. With this in mind, it is a little easier to understand why the UNIX `who` command was chosen as an example; it doesn't matter what was chosen because the application isn't what is being taught.

4.10.1 UNIX who

The UNIX `who` command lists all the users logged into a machine. In this section, two versions of an agent are developed that will travel from machine to machine, execute the UNIX `who` command on each machine, and then return to the home site and show the complete list of users to its owner. These examples are a simplistic use of an agent, but they illustrate the general structure of Internet agents, they do not require support agents at each network site, and they fit conveniently on a few pages while demonstrating most of the agent commands. As you work through these examples, you should keep in mind that the application-specific section of each agent—that is, the invocation of the UNIX `who` command—can be replaced with any desired processing.

4.10.2 Who #1

The first step in developing the examples is to install the Agent Tcl system on two or more machines (the examples work with only one machine but are somewhat boring). Detailed compilation and installation instructions are included on the CD-ROM. Once the Agent Tcl system is installed, you will have three executable files, `agentd`, `agent` and `agent-tk`. `agentd` is the agent server, `agent` is the agent interpreter, and `agent-tk` is the agent interpreter that includes the Tk toolkit. You should start the server `agentd` on each machine on which you installed the Agent Tcl system. Detailed server instructions are also included on the CD-ROM.

Once the server is running on each machine, you can execute Agent Tcl agents or any Tcl script that is fully compatible with Tcl 7.4 and Tk 4.0. Tcl scripts that require Tcl 7.5 and Tk 4.1 will not work with this version of Agent Tcl. There are three ways to execute a Tcl script using the agent interpreters. Suppose that the factorial script above is in a file called `factorial.tcl`. The first execution method is to start the agent interpreter by typing `agent` at the UNIX prompt. Then you type `source factorial.tcl` at the Tcl prompt. You will return to the Tcl prompt after the factorial script finishes executing; you can type in additional Tcl commands or type `exit` to leave the agent interpreter and return to the UNIX prompt.

The second execution method is to type `agent factorial.tcl` at the UNIX prompt; you will return to the UNIX prompt when the factorial script has finished executing. The third execution method is to turn on the UNIX execution permissions for file `factorial.tcl` and add the line:

```
#!/usr/local/bin/agent
```

at the beginning of `factorial.tcl`. This assumes that the `agent` interpreter is in directory `/usr/local/bin`; you will need to change this line if you installed `agent` in a different directory. Then you simply type `factorial.tcl` at the UNIX prompt; you will return to the UNIX prompt once the factorial script finishes executing. If the agent uses Tk, you use the same three execution methods, only with `agent-tk` rather than `agent`. Since the Agent Tcl system uses a modified Tcl interpreter, you must execute agents with either `agent` or `agent-tk`. It is impossible to execute an agent with the standard Tcl interpreters, `tclsh` and `wish`, even if you recompile them so that they include the agent extension.

Now two versions of the `who` agent are developed. The first version is text-based. It asks the user for a list of machines. Then it submits a single child agent using the `agent_submit` command. This child agent migrates through the specified machines using the `agent_jump` command, executes the UNIX `who` command on each machine, and records the users of each machine. Once the child agent finishes, it sends the complete list of users to its parent using the `agent_send` command. The parent displays the list of users and exits. Figure 4.2 illustrates the behavior of this agent.

As illustrated, the parent agent submits a child agent that migrates through a sequence of machines and executes the UNIX who command on each. Then the child sends the complete list of users to the parent for display to the user. In the specific case shown, the child agent migrates through four machines, `cosmo`, `lost-ark`, `temple-doom`, and `tuolomne`.

The Tcl code for this agent is actually quite simple. You can enter the code using any standard UNIX text editor. Once you have entered the code, you should save it in a file with extension `.tcl`. The discussion below assumes that you use the filename `who.tcl`. If you do not want to enter the code yourself, it is included on the CD-ROM in file `agent-tcl/book-examples/who.tcl`.

The Tcl code for the agent appears below. Segments of code are followed by discussion. Make sure that you do not type in the discussion as part of the agent. In addition, certain lines end with a backslash, which is the Tcl line-continuation character. There should not be any spaces or tabs after these backslashes. The first piece of code is simply a comment header.

```
#!/usr/local/bin/agent
#
# who.tcl
#
# This agent executes the "who" command on multiple machines.
# It submits a SINGLE child agent. The child jumps from
# machine to machine and executes the WHO command on each
# machine. Then the child returns the complete list of users
# to the parent for display.
```

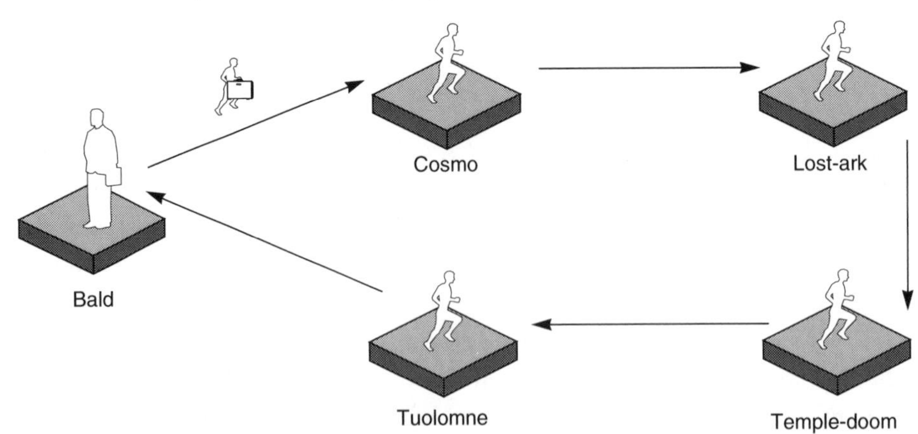

Figure 4.2 The first version of the who agent.

The first line specifies the location of the agent interpreter. This line allows you to execute the agent simply by typing who.tcl at the UNIX prompt. You will have to change this line if you installed agent in a different directory. The other lines are comments which are indicated by a single pound sign (#).

The second piece of code is the procedure that implements the *child agent.*

```
    # Procedure 'who' is the child agent that does the jumping.

proc who machines {
  global agent

    # start with an empty list

  set list ""

    # loop through the machines and jump to each

  foreach m $machines {

      # if we do not jump successfully, append an error message
      # otherwise append the list of users

    if {[catch {agent_jump $m} result]} {
      append list "$m:nunable to JUMP here ($result)nn"
    } else {
      set users [exec who]
      append list "$agent(local-server):n$usersnn"
    }
  }

    # send back the list of users and finish

  agent_send $agent(root) 0 $list
  exit
}
```

There are several important things to note about this procedure. First, the procedure takes a single argument machines which contains the list of machines that the child agent should visit. For the purposes of the examples, a Tcl list is just a string that contains one or more whitespace-separated substrings—for example, the string bald cosmo lost-ark is a Tcl list that contains three elements, bald, cosmo and lost-ark.

Second, the command global agent tells the Tcl interpreter to access the global array agent from inside the procedure; this array contains information about the location of the agent.

Third, the foreach command loops through each element in the list of machines; the variable m is set to the next machine on each iteration.

Fourth, the `agent_jump` command is used to jump onto each machine `m`. The `agent_jump` command is enclosed within a `catch` command.

Tcl commands raise *exceptions* if an error occurs; these exceptions are caught with the `catch` command. If the `agent_jump` command fails, the `catch` command catches the exception, puts the associated error message in the variable `result`, and returns 1. The *if* clause of the `if` statement is executed and the agent records an error message. If `agent_jump` succeeds, the `catch` command returns 0. The *else* clause is executed so the agent invokes the UNIX `who` command and records the list of users. Finally, once the child agent has migrated through each machine, it sends the list of users (and error messages) back to its parent using the `agent_send` command.

When agents create other agents, a parent-child hierarchy arises with a single agent at the top. The agent at the top is called the *root* agent and, in both itself and all of its descendents, its 4-element identification is found in element `root` of the `agent` array. Thus, since the parent of the child agent is also the root agent in this case, the list of users can just be sent to `agent(root)`. A current limitation of the Agent Tcl system is that it does not record the complete parent-child hierarchy. If the desire was to send the message to the parent and the parent was not a root agent, the 4-element identification of the parent would have to be explicitly recorded in an auxiliary variable before creating the child agent.

The next piece of code is the start of the parent agent. It asks for the list of machines and registers the agent with the agent server.

```
  # get the machines

puts -nonewline "Please enter the list of machines: "
gets stdin machines

  # register the agent

if {[catch {agent_begin} result]} {
  return -code error "ERROR: unable to register on
    $agent(actual-server) ($result)"
}
```

The `gets` and `puts` commands let the user enter the list of machines. The `agent_begin` command registers the agent with the server on the local machine. The `agent_begin` command is enclosed within a `catch` command in case the server is not available on the local machine for some reason (element `actual-sever` of the `agent` array always contains the name of the local machine). The agent cannot use any of the other agent commands until it successfully registers using the `agent_begin` command.

The final piece of code is the rest of the parent agent. It creates the child agent, waits for the child agent to send the message containing the list of users, and finally displays the list of users.

```
    # catch any error

if {[catch {

    # submit the child agent that does the jumping

  agent_submit $agent(local-ip) -vars machines -procs who
    -script {who $machines}

    # wait for the list of users

  agent_receive code message -blocking

    # output the list of users

  puts "nWHO'S WHO on our computersnn$message"

    # cleanup

  agent_end
} error_message]} then {

    # cleanup on error

  agent_end

    # throw the error message up to the next level

  return -code error -errorcode $errorCode
    -errorinfo $errorInfo error_message
}
```

First, the parent creates the child agent using `agent_submit`. The child agent is specified with the `-script` parameter and consists only of a call to procedure `who` with parameter `machines`. Since the child makes this call, it must have copies of procedure `who` and variable `machines`, so this procedure and variable are specified after the `-procs` and `-vars` parameters respectively. Once the child agent is created, the parent waits for the child's message using the `agent_receive` command.

The `-blocking` parameter indicates that the agent will wait until the message arrives rather than time out. Once the message arrives, the integer code is placed in variable `code` and the string is placed in variable `string`.

Finally, the `puts` command displays the list of users and the `agent_end` command ends the agent. This whole sequence is enclosed in a `catch` command in case an error occurs. The agent is now complete and can be run with any of the three methods described above. So if you type `agent who.tcl` at the UNIX prompt, you will see the request:

```
Please enter the list of machines:
```

You should type in the desired machine names with one or more spaces between names. The agent server must be running on the specified machines. As an example, if the agent were executed at Dartmouth and you entered the same machine names shown in Figure 4.2 (as well as one machine that does not exist), you might see the output:

```
Please enter the list of machines:
cosmo lost-ark xxx temple-doom tioga

WHO'S WHO on our computers

cosmo.dartmouth.edu:

lost-ark.dartmouth.edu:
lwilson     ttyq0        Apr 29 08:16
pascalb     ttyq2        Apr 29 09:11
pascalb     ttyq3        Apr 29 09:11

xxx:
unable to JUMP here (unable to get IP address of "xxx")

temple-doom.dartmouth.edu:
rgray       ttyq0        Apr 29 08:55
rgray       ttyq2        Apr 29 09:08

tioga.cs.dartmouth.edu:
rgray        ttyp2       Apr 29 09:07
```

There will be a short delay before the child agent finishes its travels and the list of users is displayed. Note that the nonexistent machine `xxx` causes no difficulties due to the `catch` command surrounding the `agent_jump` command. Detecting and handling errors when the agent is moving is no more difficult than when the agent is stationary. Uncaught errors cause the agent to terminate, although an error message will be automatically sent to the *root* agent

4.10.3 Who #2

The second version of the `who` agent expands on the first. First, it uses the Tk toolkit to display a window in which the user enters the names of the machines. Then, the agent itself jumps from machine to machine and executes the UNIX `who` command on each machine.

Once the agent has migrated through each machine, it jumps again to return to its home machine where it displays a second window that contains the results. As an additional feature, the agent leaves behind a tracker agent on the home machine; the agent communicates with the tracker agent to provide a continuous update of its current status and network location. A sample screen dump is shown in Figure 4.4. This agent is much longer so you will probably want to use the copy in `agent-tcl/book-examples/winwho.tcl` rather than typing it in yourself. All of the code should be placed in one file although logically there are two agents (the `who` agent creates the `tracker` agent just before it starts to migrate). The first piece of the `who` agent is again a comment header. The only difference is that the first line must specify the location of the `agent-tk` interpreter rather than the `agent` interpreter.

```
#!/usr/contrib/bin/agent-tk
#
# who.tk
#
# This agent executes the who command on multiple machines.
# It displays a Tk window in which the user enters a list of
# machines.  Then it jumps from machine to machine and executes
# the UNIX who command on each machine.  Finally it returns
# to the home machine and displays a Tk window that contains
# the complete list of users.  While traveling, it leaves
# behind a tracker agent; it communicates with the tracker
# agent to display continuous information about its progress.
```

The second piece of the `who` agent comprises the procedures `GetMachines` and `DisplayList`. Procedure `GetMachines` creates the window in which the user enters the machine names; this window is the top window in Figure 4.4. Procedure `DisplayList`

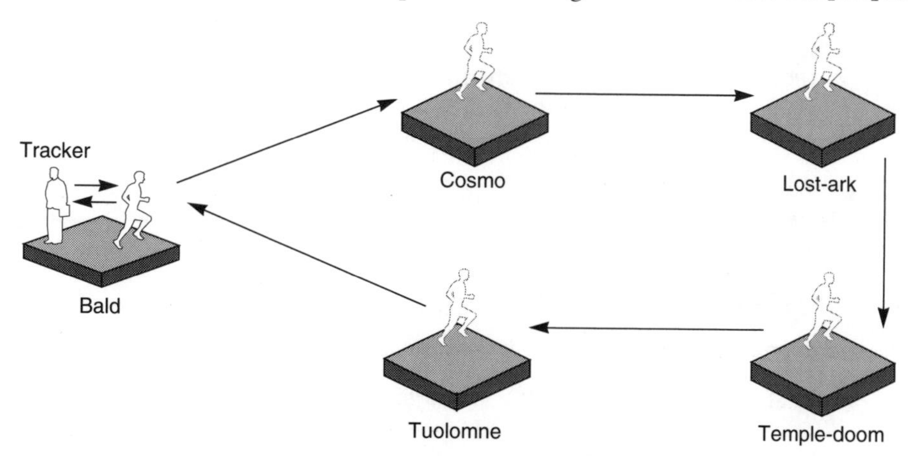

Figure 4.3 The second version of the who agent.

creates the output window in which the list of users is displayed; the output window is the bottom window in Figure 4.4.

Procedure `GetMachines` is called before the agent starts migrating; procedure `DisplayList` is called when the agent returns to the home machine with the list of users. These procedures use standard Tk commands and do not use any agent commands, so they are not described in detail. The only nonstandard commands are `main create` and `main destroy`, which create and destroy a main window for the application. The standard Tk interpreter, `wish`, automatically creates a main window. Agents, however, do not always need a main window so the command `main create` is introduced to explicitly create the main window when desired.

In addition, an agent can not migrate if it is currently displaying a window. For this reason `main destroy` is used to destroy the main window before migration. Unlike `wish`, destroying the main window does not terminate the agent. Because of the need to destroy windows before migrating—and because agents can not jump from inside a Tk event handler—Tk agents make heavy use of the `tkwait` command. The agent displays the desired interface, uses `tkwait` to stay in the event loop until the agent needs to migrate, and then destroys the interface and jumps to the next machine.

This approach imposes a useful structure on the agent and is more convenient than it might seem.

```
    # Procedure GetMachines creates the Tk window in which the
    # user enters the list of machines. It returns "OK" if the
    # user enters a list of machines and selects the "GO" button
    # It returns "FORGET" if the user selects the "FORGET" button.

proc GetMachines {} {

    # The global variable "machines" holds the list of machines
    # and the global variable "status" is either "GO" or
    # "FORGET" depending on which button the user hits. The
    # global variable "display" holds the name of the display
    # --- e.g., # "cosmo.dartmouth.edu:0".

   global display
   global machines
   global status

    # create the main window

main create -name "List of machines" -display $display

    # fill in the main window with an entry box and two buttons

     entry .entry -width 40 -relief sunken -bd 2
       -textvariable machines
     button .go -text "Go!" -command {set status GO}
```

```
    button .forget -text "Forget it!" -command {set status FORGET}
    pack .entry -side top -fill x -expand 1
    pack .go -side left -padx 3m -pady 3m -expand 1
    pack .forget -side left -padx 3m -pady 3m -expand 1
    bind .entry <Return {set status GO}
    focus .entry

      # wait for the user to fill in the entry box correctly,
      # first making sure that the "status" variable does not yet
      # exist

  catch {unset status}

  while {![info exists status]} {

      # wait for the user to hit a button

    tkwait variable status

      # if the user hit button "GO", see if the entry box is
      # filled in

    if {($status == "GO") && ([string trim $machines] == "")} {
          tk_dialog .t "No machine!"
            "You must enter at least one machine name!" error 0 OK
          unset status
      }
    }

      # return the status --- e.g., "GO" or "FORGET" -- but first
      # destroy the window

    main destroy
    return $status
  }

      # Procedure DisplayList creates the window in which the list
      # of users is displayed. The "users" argument contains the
      # list of users.

  proc DisplayList users {

      # The global variable "display" contains the name of the
      # display and the global variable "status" will be set to
      # DONE when the user finishes looking at the results.

    global display
    global status

      # create the main window

    main create -name "WHO'S WHERE?" -display $display
```

```
    # make the placeholder frames

frame .top -relief raised -bd 1
frame .bot -relief raised -bd 1
pack .bot -side bottom -fill both
pack .top -side bottom -fill both -expand 1

    # make a text box that will hold the list of users

text .text -relief raised -bd 2 -width 60
  -yscrollcommand ".scroll set"
scrollbar .scroll -command ".text yview"
pack .scroll -in .top -side right -fill y
pack .text -in .top -side left -fill both -expand 1

    # make the "DONE" button

button .done -text "Done!" -command {set status DONE}
pack .done -in .bot -side left -expand 1 -padx 3m -pady 2m

    # fill in the text area

.text delete 1.0 end
.text insert end $users

    # wait for the user to finish looking at the results, first
    # making sure that the "status" variable does not yet exist

report "Done! You should see the results window."
catch {unset status}
tkwait variable status
main destroy
}
```

The next piece of the who agent is actually the tracker agent that displays the progress of the who agent through the network. The who agent uses the `agent_event` command to send tagged messages back to the tracker. Rather than explicitly receiving these messages with the `agent_getevent` command, the tracker uses the `mask` command to establish two message handlers. These handlers are automatically called when a tagged message arrives.

Procedure `messageHandler` is automatically called if the message tag is `MESSAGE`. The `source` parameter is filled in with the 4-element identification of the sender; the `tag` parameter is filled in with the message tag; and the `string` parameter is filled in with the message string. Similarly procedure `errorHandler` is called if the message tag is `ERROR`. Procedure `Tracker` is the main body of the tracker agent. It creates a simple text window, establishes the two message handlers using the `mask` command, and calls `tkwait` to sit in the event loop.

The two handlers are automatically called whenever a message arrives and simply insert the status information into the text window. This text window is the middle window in Figure 4.4. The tracker agent illustrates that agents can use the Tk event model effectively. In fact Tk agents should almost always establish event handlers for incoming messages; otherwise the agent will not respond to user events while it sits at an `agent_receive` or `agent_getevent` command (or it will have to continuously poll).

Procedure `LeaveTracker` actually starts up the tracker agent using `agent_submit`; it is called by the who agent just before the who agent starts migrating. The procedure returns the 4-element identification of the tracker so that the who agent knows where to send its status messages.

```
    # Procedure errorHandler, messageHandler and Tracker make up
    # the tracker agent. Procedure LeaveTracker starts the
    # tracker agent and returns either the 4-element id of the
    # tracker or the string "FAILED".

proc messageHandler {source tag string} {
    .text insert end "$stringn"
}


proc errorHandler {source tag string} {
    .text insert end "nERROR: $stringnn"
bell
}

proc Tracker {} {

    # The global variable "display" holds the name of the
    # display.   The global variable "status" will be set to
    # DONE when the user decides to exit. The global array
    # "mask" --- which is available inside every agent ---
    # specifies event handlers.

  global display
  global status
  global mask

    # create the tracker window

    main create -name "Tracker agent" -display $display

    # make the placeholder frames

    frame .top -relief raised -bd 1
    frame .bot -relief raised -bd 1
    pack .bot -side bottom -fill both
    pack .top -side bottom -fill both -expand 1
```

```
     # make a text box that will hold the list of users

  text .text -relief raised -bd 2 -width 60
  -yscrollcommand ".scroll set"
  scrollbar .scroll -command ".text yview"
  pack .scroll -in .top -side right -fill y
  pack .text -in .top -side left -fill both -expand 1

     # make the "DONE" button

  button .done -text "Done!" -command {set status DONE}
  pack .done -in .bot -side left -expand 1 -padx 3m -pady 2m

     # turn on the event handlers

  mask add $mask(event) "ANY -tag MESSAGE
     -handler messageHandler"
  mask add $mask(event) "ANY -tag ERROR -handler errorHandler"

     # wait for the user to finish looking at the results, first
     # making sure that the variable "status" does not yet exist

  catch {unset status}
  tkwait variable status
  main destroy
}

proc LeaveTracker {} {

   global agent
   global display

      # try to submit the tracker agent

   if {[catch {

      set tracker [
         agent_submit $agent(local-ip) -vars display
            -procs errorHandler messageHandler Tracker
            -script {Tracker; exit}
      ]

   } result]} {

      set tracker FAILED

   }

   return $tracker
}
```

The next piece of the who agent is procedure `who`, which routes the agent through the specified machines using `agent_jump` and executes the UNIX `who` command on each. This procedure is almost the same as the who procedure from the first version. The only difference is that it reports its current location and status to the tracker agent by calling the report and reportError procedures.

These two procedures use agent_event to send a tagged message back to the tracker. When the tracker receives the tagged message, either procedure messageHandler or procedure errorHandler is automatically called, and the status information is inserted into the tracker window.

```
        # Procedure who executes the UNIX "who" command on each
        # machine. Procedure report sends normal information back to
        # the tracker agent whereas Procedure reportError sends error
        # information back to the tracker agent.

    proc report message {

        # The global variable "tracker" holds the 4-element id of
        # the tracker agent.

      global tracker

        # send the message, ignoring errors

      catch {
         agent_event $tracker MESSAGE $message
       }
     }

    proc reportError error {

        # The global variable "tracker" holds the 4-element id of
        # the tracker agent.

      global tracker

        # send the message, ignoring errors

      catch {
         agent_event $tracker ERROR $error
       }
     }

    proc who machines {

      global agent
      global tracker

        # start with an empty list

      set list ""

        # jump from machine to machine
```

```
    foreach m $machines {

        # if we do not jump successfully, append an error message
        # otherwise append the list of users

        if {[catch "agent_jump $m" result]} {
          reportError "Failed to jump to machine $m ($result)"
          append list
            "$m:nunable to JUMP to this machine ($result)nn"
        } else {
          report "Jumped to machine $agent(actual-server)"
          set users [exec who]
          append list "$agent(local-server):n$usersnn"
      }
    }

    return $list
  }
```

The last piece of the who agent simply calls the procedures above. First, the who agent calls procedure `GetMachines` to get the machine names from the user; the machine names are stored in the global variable `machines`. Once the machine names have been obtained, the agent calls `agent_begin` to register the agent with the local agent server, and then calls procedure `LeaveTracker` to start up the tracker agent. Then the who agent jumps through the specified machines by calling procedure `who`; procedure `who` returns the list of users.

Once procedure `who` is finished, the agent calls `agent_jump` one more time to return home. Once the agent is home, it calls procedure `DisplayList` to show the list of users in an output window. Finally the agent calls `agent_end` and exits.

```
  # remember the display

if {![info exists env(DISPLAY)]} {
  set display ":0"
} else {
  set display $env(DISPLAY)
}

  # get the list of machines

if {[GetMachines] == "FORGET"} {
  exit
}

  # register the agent with an agent server and remember the
  # home machine

if {[catch {agent_begin} result]} {
```

```
    puts "Unable to register on $agent(actual-server) ($result)"
    exit
}

set home $agent(local-ip)

    # try to leave behind the tracker agent

set tracker [LeaveTracker]

if {$tracker == "FAILED"} {
    puts "Unable to leave behind the tracker agent!"
    exit
}

    # jump from machine to machine, executing the "who" command on
    # each machine, and then jump back home

set users [who $machines]
agent_jump $home

    # display the results

DisplayList $users

    # done

exit
```

The agent is now complete. It can be run with any of the three methods discussed above except that you must use `agent-tk` rather than `agent`. One important note is that, if you followed the installation instructions carefully (which is highly recommended), an agent will start running under a special user ID as soon as it jumps for the first time. On most UNIX machines, you will need to use the `xhost` command (or equivalent) to allow this special user ID to create windows on your screen; otherwise the agent will not be able to create the output and tracker windows.

The reference documentation for your UNIX machine will have more details about screen access. Once the agent starts executing, you will first see the entry form where you enter the names of the machines. Once you hit `GO!` to send the agent on its way, the entry form will disappear, and the tracker window will appear. Lines will appear in the tracker window one at a time as the who agent makes its ways through the network and reports back its current location. Finally the who agent will return and the output window will appear showing the list of users. A sample run is shown in Figure 4.4; the machine names are the same as were used before.

Although these two versions of the who agent perform a simple task, they use most of the agent commands and can serve as building blocks for more complex agents. There is no reason for the agent to be self-contained, for example. There might be service agents on each machine with which the agent communicates as it migrates. These service agents should be given well-known names with the `agent_name` command so that client agents can communicate with them easily.

In one of the information-retrieval applications, for example, there is an agent named `TechReports` on each machine which provides a low-level search interface to a collection of technical reports. Agents, migrating from collection to collection, combine the low-level search primitives into complex queries.

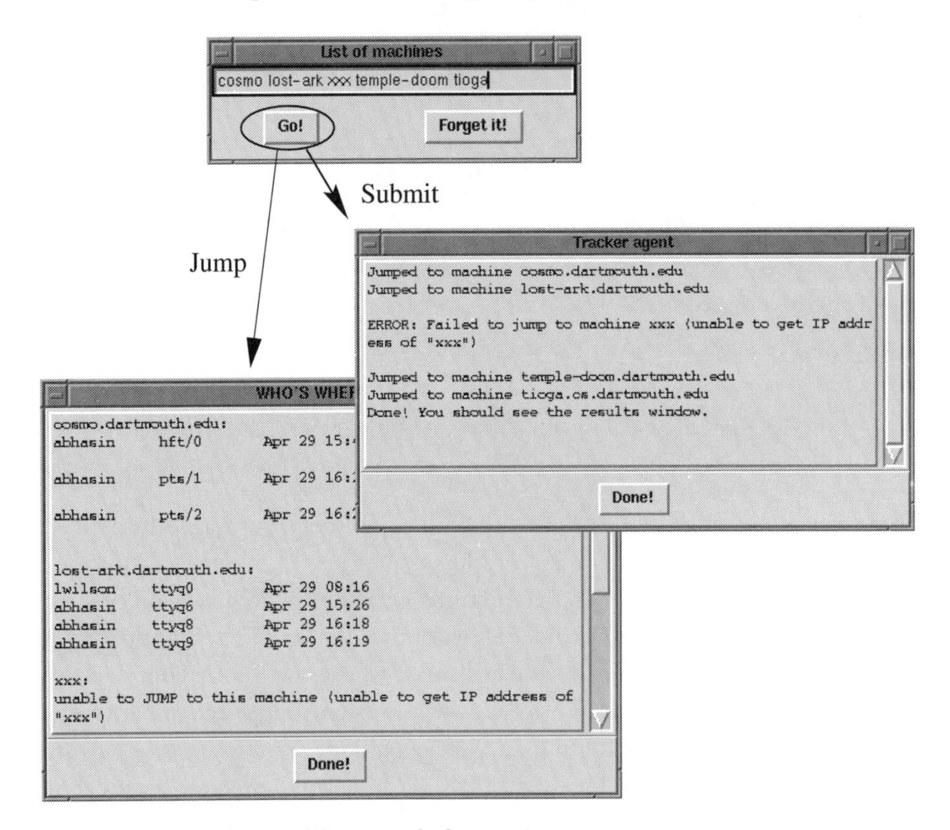

Figure 4.4 A sample run of the second who agent.

One area of difficulty for new agent programmers is debugging a moving agent. Agent Tcl does not include a visual debugger, but several debugging strategies are discussed in the documentation, and each is reasonably effective.

One of the best is illustrated by the second who agent—that is, a moving agent continually reports its status to some specified tracker agent. To report Tcl exceptions, the main body of the agent can be surrounded with a `catch` command; if this `catch` command catches an error, the complete error message can be sent to the tracker (as well as the error location since Tcl maintains a stack trace in the global variable `errorInfo`). Once the agent is debugged, the tracking code can be removed.

4.11 Pros, cons, and advantages

Agent Tcl involves several trade-offs. Like Ara, Agent Tcl uses the simple scripting language, Tcl, as the main agent language. Other Internet agent systems such as Telescript and Java require the programmer to use a complex, object-oriented language even for simple agents. In addition, few systems other than Visual Obliq provide a graphical toolkit that is as high-level and flexible as the Tk toolkit. Agent Tcl, therefore, allows much more rapid development of small- to medium-sized applications.

Tcl, however, is slow compared to other scripting languages and is much slower than interpreted bytecodes and native machine code. In addition, Tcl provides no code modularization aside from procedures. Agent Tcl, therefore, can not be used in speed-critical or large applications. Searching a large, distributed collection of numerical data or performing intensive mathematical calculations, for example, would be intolerably slow without at least some low-level support at each site. Developing a mobile, full-featured word processor would involve too much Tcl code to be practical (although the application would potentially be fast enough with careful Tk programming). Java, Telescript, and Ara, which compile their agents into interpreted bytecodes, are the only reasonable choice for such applications, although even these systems would be too slow for such things as distributed scientific computing.

Agent Tcl provides simple, flexible migration and communication primitives. Like Telescript, Agent Tcl provides the *jump* primitive, which captures the complete state of the agent and transparently sends the state to the destination machine. Tacoma, on the other hand, requires the programmer to explicitly collect state information in a briefcase and then submit this briefcase along with the migrating agent; the agent starts execution from the beginning and must use the information in the briefcase to determine which task to perform next.

Both approaches are equally powerful, but the jump primitive is more convenient. There is the potential to overuse jump and write hard-to-understand code—for example, calling a procedure might unexpectedly move the agent to a new location because there is a *jump* buried in the code. This problem is much less severe than the historic *goto* problem, however, since there are no unexpected changes in control flow, and it appears that the problem is not severe enough to outweigh the convenience.

Agent Tcl's communication primitives hide all the transmission details but are low-level enough to efficiently support a range of higher-level communication services. Some systems, such as SodaBot, provide a specific high-level communication paradigm (e.g., actor-based, declarative logic, and so forth.) that is inappropriate for many applications. The programmer is either locked into this paradigm or forced to communicate outside of the agent framework.

Agent Tcl's communication primitives have two drawbacks. First, if a higher-level communication protocol is desired, it must be implemented on top of the low-level primitives. Second, there is no common language that every agent understands. The flexibility of low-level primitives outweighs these drawbacks. It is expected that several standard, high-level communication protocols will eventually be provided as part of the Agent Tcl system; RPC and dialog-based mechanisms have already been implemented but are not included on the CD-ROM.

In addition, agents might be required to understand one simple, common protocol for exchanging status information, but allow them to use any other protocol that they saw fit.

Agent Tcl's main weakness is that it does not provide the features of more mature systems. Agent Tcl lacks the visual debugging tools of Java and Telescript, although a simple visual debugger exists and is being tested. Similarly, the version of Agent Tcl on the CD-ROM does not provide the security features of Telescript. Telescript authenticates all incoming agents and assigns access restrictions based on this authentication.

The development version of Agent Tcl, however, does exactly this using PGP and Safe-Tcl (the development version will be released in early 1998). Agent Tcl's security model, in which resource managers assign access restrictions based on the agent's identification, is simpler than the Telescript model. Telescript agents communicate by exchanging references to each other's objects. Handling the security problems that arise when agents call into each other's objects requires awkward class syntax and paranoia programming on the part of the agent programmer.

Exchanging object references has the additional drawback of making it difficult to include new languages in a Telescript system. One of the main research areas of Agent Tcl is to expand on existing security mechanisms so that the system protects agents and groups of machines in addition to individual machines.

The version of Agent Tcl on the CD-ROM also does not include direct support for mobile computing; Telescript does provide such support. A flexible system of support agents for mobile computing has been implemented, however, and is successfully using these agents in several applications. Although these fault-tolerance mechanisms are not incompatible with Agent Tcl, addition of them is not planned as part of the research.

Agent Tcl does not yet support multiple languages. Work on incorporating Java, however, is progressing well. Finally, from an architectural standpoint, Agent Tcl is inefficient since the server and each agent run as separate processes, rather than in an integrated execution environment such as Ara or Telescript. There is no plan to change this in the near future.

Agent Tcl, therefore, is best-suited for experimentation with Internet agent ideas and for the development of small- to medium-sized applications in which at least some low-level support is available at each site. Agent Tcl agents combine the low-level services at each site into complex operations, coordinate their efforts with other agents, and handle unexpected error conditions. The flexibility of Agent Tcl allows such agents to be developed rapidly.

chapter 5

Agents for remote access

CO-AUTHORED BY HOLGER PEINE

The basic idea behind the Agents for Remote Access (Ara) system is the development of a platform for agents able to move freely and easily at their own pace and without interfering with their execution, utilizing various existing programming languages and even existing programs, independent of the operating systems of the participating machines.

This chapter is structured as follows: The initial section introduces the basic concepts of Ara, such as languages, agents and mobility. The second section demonstrates how common problems of networked computing can be solved using these concepts. This is followed by a section explaining the individual features and facilities of Ara as they are presented to the programmer. These are put to use in an example for searching the World Wide Web (WWW). The subsequent section discusses selected aspects of the Ara system architecture to deepen the understanding of the system's capabilities and shows how to extend it with other programming languages. The chapter concludes with a discussion of different approaches to mobile agent systems and future developments of Ara.

The example developed in Ara is an agent that travels around to Web sites looking for "interesting" information and sending it back to the user. The agent directs its travels based on the information which its finds; it does not have a pre-defined path for its search.

5.1 *Overview*

Mobile agents are a powerful new concept in networked computing with far-reaching implications, many of which are discussed in this book. Still, the basic idea is quite simple: Give programs the ability to move. It seems natural, therefore, to add mobility to the large and well-developed world of programming, rather than attempt to build a new realm of mobile programming. Mobility should be integrated as comfortably and unintrusively as possible with existing programming concepts—algorithms, languages, programs, and operating systems. This is the basic idea of the Agents for Remote Access (Ara) system: a platform for agents able to move freely and easily without interfering with their execution, utilizing various existing programming languages and existing programs, independent of the operating systems of the participating machines. Complementing this, the system provides facilities for the specific requirements of mobile agents in real applications, security concerns being the most prominent.

This chapter will present Ara and its specific approach to Internet agents (or mobile agents, as they are termed in Ara), explain the concepts and features of the system, and demonstrate how they are used to program mobile agents to solve real-world problems. Ara is an example of middleware, situated between specific applications (e.g. an airline booking system) and the underlying operating system. It provides the system-level facilities to execute and move programs (agents), let them interact and access their host system, all in a portable and secure manner.

The actual application (the real work) is of no interest to the system, but programmed in the behavior of the agents, expressed using various programming languages on top of a common run-time core. The system should be seen as a broker between visiting agents and the host system, providing the agents with access to local services and the host with control over the agents. And like any good broker, its policy is not to interfere, but to let both parties do their business as they see fit, yet oversee that they play by the rules of honest business.

A remark is in order here concerning the maturity and completeness of the Ara system at the time of this writing. The system is in active development; while the basic concepts are resolved and the system is sufficiently worked out for useful applications (see section 5.3, "Mobile agent applications with Ara"), many of the more advanced features are not yet implemented, and some of them are still in the process of definition. This will be indicated in the presentation where appropriate.

The rest of this chapter is structured as follows: the initial section introduces the basic concepts of Ara, such as languages, agents, and mobility. The second section demonstrates how common problems of networked computing can be solved using these concepts. This is followed by a section explaining the individual features and facilities of Ara as they are presented to the programmer. These are put to use in an example for searching the World Wide Web (WWW). The subsequent section discusses selected aspects of the Ara system architecture to deepen the understanding of the system's capabilities and shows how to extend it with other programming languages. The chapter concludes with a discussion of different approaches to mobile agent systems and future developments of Ara.

5.2 An overview of Ara

From a high-level view, Ara consists of agents moving between places, where they use certain services to do their job. The fundamental concepts of agents, motion, places, and service access in Ara will be explained in this section. As agents are written in some programming language, the role of languages within the Ara system will also be of interest.

5.2.1 Agents, languages, and the Ara system

There has been, and no doubt will continue to be, a broad discussion about what constitutes an agent, as opposed to a general program. This discussion has been both controversial and confusing, and apparently has not yet reached a consensus beyond a few buzzwords, autonomous being the most prominent. Ara's policy, stressing application independence, is not to take sides in that discussion. Instead, the Ara notion of an agent is simply derived from the more clearly defined notion of a mobile agent.[1]

In Ara a mobile agent is a program with the ability to move during execution. Besides mobility, there is nothing new to a mobile agent. Of course, this could be called an intentionally misleading statement, since the very mobility of a program across different machines has truly far-reaching consequences. This affects the design of the runtime environment for such code, most notably portability and security concerns.

Portability and security are fundamental to mobile agent systems. Portability is an issue because mobile agents should be able to move in heterogeneous networks (between machines with different operating systems and hardware architectures) to be really useful. Security is important because the agent's host effectively hands over control to a foreign program of unknown effect.

There is also the reverse problem of the agent's security against undue actions of the host, for example, spying on the agent's content. There is, however, no general solution for this problem except several individual aspects. Most existing mobile agent systems, while differing considerably in practice, use the same basic solution for portability and security. They do not run the agents on the real machine, but on a virtual one, usually an interpreter and a run-time system, which hides the details of the host system architecture and confines the actions of the agents to that restricted environment.

The concept of a dedicated execution environment, providing a secure and portable set of services to access the host system and possibly other agents, enables agents to move in heterogeneous networks and permits a fine-grained control of the executing agent without depending on the hardware platform.

This is the approach adopted in Ara. Mobile agents are programmed in an interpreted language and executed within an interpreter for this language, using a special run-time system for agents, called the core in Ara terms. However, the relation between core and interpreter is characteristic for Ara: isolate the language-specific issues (how to capture the Tcl-specific state of an agent programmed in the Tcl programming language) in the interpreter, while concentrating all language-independent functionality (how to capture the general state of an agent and use that for moving the agent) in the core.

This separation of concerns makes it possible to employ several interpreters for different programming languages together on top of the common, generic core. The core deals with general agents only, making its services uniformly available to all agents

---

1 Although, even agents which cannot really move, such as stationary servers, are subsumed under the term agent in Ara. This is simply because it is convenient to have a common term for all active entities in the system.

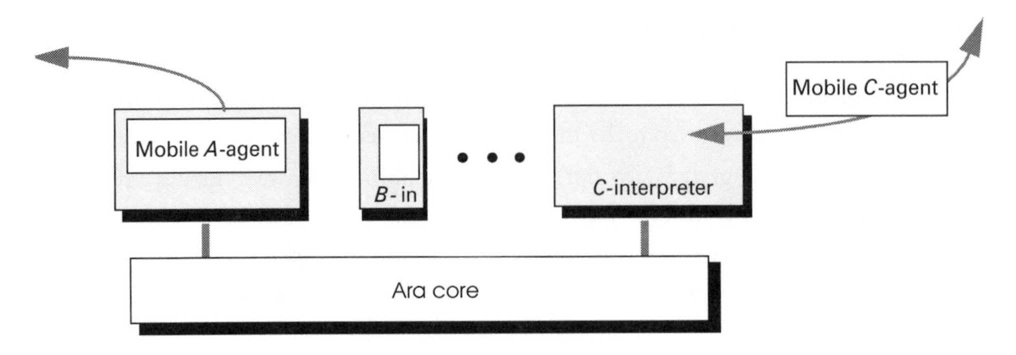

Figure 5.1 Agents, interpreters and the core

regardless of their respective interpreter languages. Although this matter will be treated more thoroughly in section 5.6.1, "Processes and internal architecture," it should be stated here that the complete Ara system of agents, interpreters and core runs as a single application process on top of an unmodified host operating system. Figure 5.1 shows this relation of agents, core, and interpreters for languages called A, B, and C.

While Ara stresses the independence of its concepts for mobile agents from specific programming languages, the choice of languages is not irrelevant. At the time of this writing, interpreters for the Tcl and C/C++ programming languages have been adapted to the Ara core, and the Java language will be added next. Tcl and C/C++ differ considerably in complexity, run-time efficiency, and development expense.

Tcl, a popular scripting language, offers low development expense and a compact, embeddable, and freely available interpreter, which have made it the language of choice of several mobile code systems.

C/C++ (to be referred to simply as C from here on) is superior regarding the available abstractions, the run-time efficiency, and the interoperability with existing software. The reader may wonder that C is used as an interpreted language here, since C as a typical compiled language is not directly suitable for interpretation.

It was considered important to utilize the most widely used programming language for mobile agents, saving the effort of adopting a new language and enabling the reuse of an enormous base of software. Accordingly, a dedicated interpreter for C was developed, specially adapted to mobile code. This interpreter works by precompilation to the portable MACE bytecode, which is subsequently interpreted with remarkable performance by the MACE interpreter. At the time of this writing, the MACE interpreter is not ready for distribution yet but it will be included in the next Ara release.

The rationale for interpretation was support for portability and security, especially the portability of a live agent's execution state (see section 5.6.4). While there seems to be no viable alternative to interpretation when the full mobile agent functionality is desired, compiled agents were integrated into the Ara system as an efficient alternative for cases where certain security and portability requirements can be sacrificed.

Such compiled agents cannot normally move to other machines, as they generally exist in a machine-dependent form only, and security cannot totally be warranted. In many respects, however, they behave like their interpreted siblings at a substantially increased speed. Compiled agents are usually employed for services resident on a local system.

Ara thus offers the option to choose a programming language, instead of requiring all applications to be written in one prescribed language. A prerequisite for this is that the desired language interpreter has been adapted to Ara. However, this adaption is well-defined and straightforward on the part of Ara (see section 5.6.6, "Adaption of further programming languages to Ara").

The programming examples in this chapter will use Tcl most of the time since it is more concise than C. C examples will be given from time to time to aid the distinction between language and concept.

5.2.2 The life of an agent

What makes an Ara agent different from conventional programs is its characteristic use of functions provided by the core for agent actions, control, and interaction. An active agent in Ara is a process; a self-contained activity with its own state and progress. This is both a natural and powerful form for agents, which are designed to be autonomous. Creation and deletion of new agents, respectively agent processes, are among the most basic functions offered by the core. Each agent is assigned a unique, immutable name on creation. Newly created agents will execute their program in parallel with other agents active on the same system. They can clone themselves; suspend, resume, and terminate their operation; sleep for some time; or be likewise acted upon by other agents, provided the necessary access rights. These agent control functions can be used to form a team of agents working on a common task.

Although many useful tasks can be performed by a single mobile agent using the functions offered by the local host, considerable flexibility and new kinds of applications are gained through interaction between agents. Agents can talk to each other, exchange information, offer and request services from one another, and even negotiate and trade. This concept is also useful for structuring a host environment, as higher-level services can be offered through system agents. This is more modular and flexible than offering the services through a static function-call interface.

It can be argued whether agent communication should be remote or restricted to agents at the same place. Considering that one of the main motivations for mobile agents was to avoid remote communication in the first place, Ara opted for local agent interaction. This is not to say that agents should be barred from network access (which depends on the policy of the host system interface, see section 5.2.4). Rather, the system encourages local agent interaction.

There are various options for such an interaction scheme, including disk files, shared memory areas (blackboards), direct message exchange, or special procedure calls. Each entails different ways of access and addressing.

Ara chose a variant of message exchange between agents, allowing client/server style interaction. The core provides the concept of a service point for this interaction. This is a meeting point with a well-known name where agents located at the same place can interact as clients and servers. An agent announces the service point, thereby assuming the role of a server agent, whereupon client agents may meet it.

After a successful meeting, a client can submit service requests to the service point, and the server can fetch those and reply to them as it sees fit. Each request is marked with the name of the client agent which the server may use in choosing a reply. For example, a service point might advertise access to a database as shown in figure 5.2, and clients may submit queries, which are fetched by servers to be inspected, perhaps preprocessed, billed, or translated before being passed to the database management system. The query results are then passed back to the client as the reply to the service request.

Service requests and replies are implemented as messages with content and meaning entirely up to the participants. Many agents can meet at a service point. An agent can play the role of both client and server at different service points at the same time.

Agents on the move are programs running out of the sight and reach of their creator. It seems wise, therefore, to have a means of setting global limits to their actions and safe-guarding against unwanted effects like circling through the net in endless loops. The potential danger of agents getting out of control becomes clearer when considering that certain hosts might charge visiting agents for the resources consumed during their stay such as execution time.

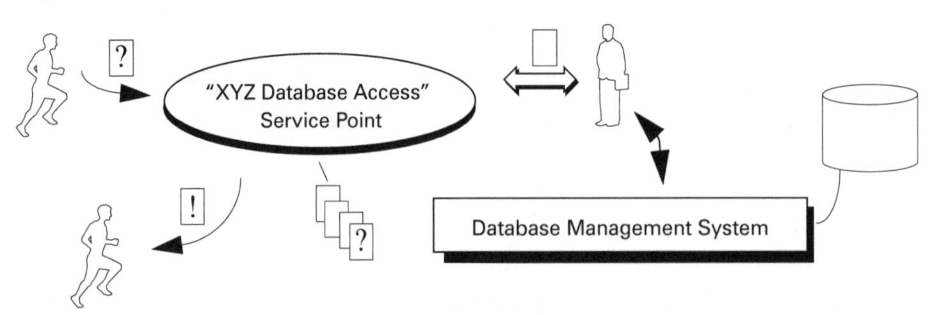

Figure 5.2 Client and server agents at a service point

The hosts, on the other hand, have an even stronger interest in setting limits to visiting agents running on their system in order to prevent overuse of host resources and to enforce resource agreements.

For the purpose of resource limitation, Ara agents are equipped with resource accounts called allowances. An allowance for a resource, such as memory, records the amount the agent is currently allowed to consume of this resource in the future. When the agent consumes some of the resource, its allowance is updated accordingly by the system.

The allowance is likewise adjusted when the agent returns some amount of a reclaimable resource. The system ensures that an agent can never overstep its allowance. In the simplest case, an agent is given an initial allowance for its own perusal at the time of agent creation. A group of agents may also share a common allowance, each consuming from it according to their own policy. Agents may inquire about their current allowance at any time and may transfer it among each other. Such transfers can be used as an organizational measure in a group of cooperating agents, but also for trading resources between buyers and sellers.

Most of the time allowances are used by the creator of an agent to bound its global range of action, and by a receiving place to limit the agent's local resource consumption. In the latter case, the receiving place temporarily imposes a local allowance on an arriving agent, further restricting the agent's global allowance set by its creator, as sketched in Figure 5.3. Additionally, the agent itself may specify a desired local allowance when moving to a place (default for this is the full global allowance). On leaving the place, any local allowance is released and the global allowance again takes effect.

Mobile agents lead a more dangerous life than ordinary, stationary programs. When moving through a large network such as the Internet they have to be prepared for all kinds of accidents happening. Agents may get blocked on the way, they may fall prey to unexpected software faults, or their current host machine may simply crash, burying the agent with it. Some of these dangers can be countered by specific measures, but there is no general protection against an agent dying unexpectedly on its itinerary.

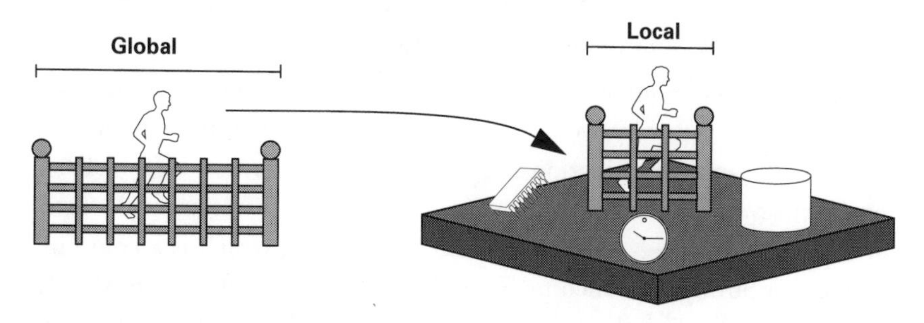

Figure 5.3 Restricting an Agent by a Local Resource Allowance

Rather than trying to anticipate all potential pitfalls, Ara offers a simple means of recovery from such accidents: An agent can create a checkpoint, a complete record of its current internal state, at any time in its execution. Checkpoints are stored on persistent media (usually a disk), and can be used at a later time to restore the checkpointed agent. On restoration, the agent resumes from the exact state before the checkpoint. The obvious application for this scheme is for an agent to leave a checkpoint behind as a back-up copy before undertaking a risky operation such as moving to unknown machines. In the event of an accident happening to the agent, it can be restored from the checkpoint and take appropriate recovery measures. Of course, the checkpointing mechanism may also be used for agents simply wishing to idle for a longer time at low-resource requirements, or for working with canned agents. Checkpointing and restoring should be used with care since restoring an agent which is indeed still alive on a remote system would produce two copies of the same agent, requiring explicit treatment.

Having covered basic agent handling, agent interaction, limitation, and recovery, the one essential aspect of mobile agents, mobility, has only casually been treated. It is time now to take a closer look at this important aspect of Ara.

5.2.3 Agent mobility: going from place to place

Moving a program between computers can mean different things. In the simplest case, the program is transported to its destination site prior to program start, and then run to completion there. It might seem that calling this a mobile agent system would be stretching the notion too far (note that this could be achieved by carrying a diskette with the program to the destination computer), but if the receiving system is aware of the foreign nature of the program and runs it in an appropriate environment, this qualifies as a mobile code system. The Java language environment is an example of such a mobile code environment.

From a point of view of mobility, the characteristic property of such systems is the restriction that a running program cannot move any further.

However, mobile agents frequently need the ability not only to move from their origin to a destination site, but to move to further sites to fulfill requirements which cannot be satisfied by the initial destination site. This might be the case when collecting information from several sources, when the final destination is not known beforehand, or when a site has to be left because it is shutting down, which is a common situation in mobile computing. For such purposes it is no longer sufficient to simply transport the program, (re)starting the agent in its initial state. Rather, the agent needs to carry additional information about its experience so far—technically speaking, about its execution state. Moving a running program, that is, a live process of code and state, is usually termed *migration* in systems programming.[2]

Ara agents can migrate at any point in their execution, simply by using a special core call, named `ara_go` in Ara's Tcl interface:[3]

```
ara_agent {ara_go moira; puts "Hello, world!"}
```

The above code creates a new agent, giving it a Tcl program (enclosed in braces) to execute. The agent will first move to a place named `moira` (a machine name, in this case) and then print the greeting message there. The `ara_go` instruction is all the programmer needs to know about migration—the system ensures that the agent is moved in whole to the place indicated and resumed exactly where it left off, directly after the `ara_go` instruction.

The complexity of extracting the complete agent from the local system, getting it to another machine and reinstalling it there, is thus hidden in a single instruction. This allows the programmer to concentrate on the application rather than on the technical details of communication. Furthermore, the act of migration does not affect the agent's flow of execution nor its set of data, allowing the programmer to make the agent migrate whenever needed without preparation or reinstallation measures, as illustrated in the following example:

```
set previousPlace [ara_here]
foreach place {moira kismet fatum} {
   ara_go $place
   puts "Hello at $place, I'm coming from $previousPlace!"
   set previousPlace $place
}
```

This agent visits the places `moira`, `kismet`, and `fatum` in turn, each time printing the place it came from.[4] The migrations are embedded in its flow of execution (a `foreach` loop in this case) without interfering with its operation, and all its data (`place` and `previousPlace` in this case) remains available. This concept of non-interfering migration is termed orthogonal migration, in contrast to variants which do affect further execution.

---

2 Note, however, that process migration, when used in operating systems context, nearly always refers to processes in tightly coupled homogeneous systems (e.g. a workstation cluster on a local network) without security and portability problems. The two concepts are clearly related, but their focus is very different.

3 The same effect could be achieved in C by calling a C function `Ara_Go()` and so on.

4 `$place` is the value of the variable `place` in Tcl syntax; `[here]` is the result of the `here` command (which returns the name of local place).

Note that while the internal state of a moving agent is transferred unchanged, this is not possible for its external state, that is, its relations to other system objects and resources like service points, files, or windows. Such objects are not mobile and have to be left behind when moving to another machine. It would be theoretically possible to add a software layer to such stationary resources making them appear mobile, effectively creating a distributed operating system. However, the complex protocols and tight coupling involved with this approach do not seem well adapted to the low-bandwidth and heterogeneous networks targeted by mobile agents.

Ara agents move between places, which is both a suggestive association to physical location and a concept of the Ara architecture. Places are virtual locations within an Ara system which are running on a certain machine; places could thus be said to refine physical location. More importantly however, an Ara place groups together logically related services and agents, structuring the offerings of a system.

Service points, for instance, are always tied to a place and can be accessed only by agents currently staying at that place. In fact, an agent is always staying at some place. Places have names which uniquely identify them and which can be specified as the destination of a migration. In practice, a place might be run by an individual, an organization, or a company presenting its services.

Figure 5.4 illustrates the relation of systems, places and service points.

In addition to structuring, places also exercise control over the agents they admit. First of all, a place has the authority to decide whether a specific arriving agent is really admitted, using the agent's name.[5] An agent which has been denied access to its desired destination place is forced to go back to its source place to discover the failure. Thus,

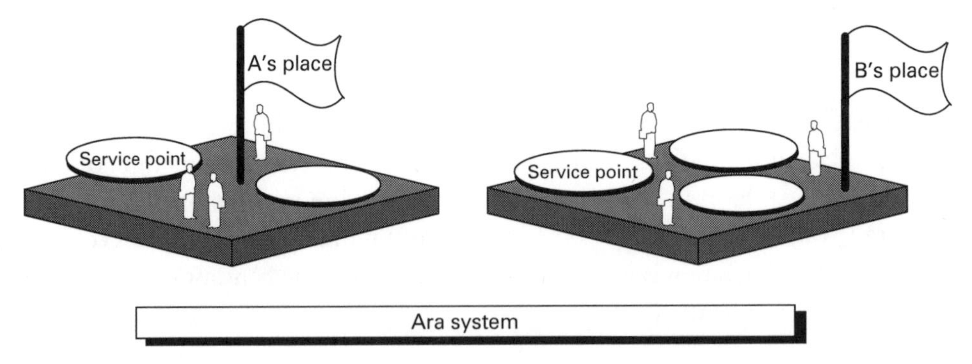

Figure 5.4 Two places with service points and agents on one system

5 If the applicant is a compiled agent, the place may also make admittance depend on the set of external functions used by that agent as an additional security measure (see section 5.4.9, "Compiled agents" for an explanation).

places play a central role in the security policy of an Ara system. In fact, each place can implement its own security policy to a large extent, requiring that arriving agents be authenticated (that is, as soon as authentication has been added to Ara) or that they have not passed through mistrusted places in their path.

Even if a place decides to admit an agent, it may impose a restrictive local allowance on it, as described in Section 5.2.2, "The life of an agent," limiting its potential actions while staying at the place. When the agent resumes, it may check its effective allowance, discovering to what extent the place has honored its desires. This enables the agent to decide on its own what to do if it finds the local allowance insufficient, rather than using a fixed negotiation mechanism built into the migration procedure.

In the current Ara implementation, places are not truly distinct objects in the system architecture. Instead, there is one implicit default place provided per system. Accordingly, place names currently reduce more or less to machine names.[6] The default place has a fixed behavior of admitting all arriving agents and does not impose local allowances on them; the agent is accepted subject to its global allowance. The next Ara release will restrict this generous policy, as well as provide the means to create new places with application-specific behavior.

5.2.4 Accessing the host system

So far, the facilities of Ara have been described at the level of (mobile) agents. Apart from that, agents need to perform physical input and output as well, accessing the user interface, the file system, external applications, and the network interface. In short, agents need access to the host system. Certainly the operating system of the host machine provides just that, but bearing in mind the increased security concerns inherent in access for mobile agents, such access has to be controllable in a more fine-grained way than is common with local programs. Additionally, the desired portability of agent programs calls for a somewhat higher level of access than normal.

In keeping with Ara's basic idea of adding mobility to existing concepts, the Ara host interface looks similar to the programming interface of common operating systems, restricted and simplified as necessary. Unfortunately, at the time of this writing, the host interface is still in development and not ready for distribution, parts being implemented, and others being designed. Therefore, this section will give only an overview of the forthcoming host interface as sketched in figure 5.5.

---

6 Actually, a place name currently designates a specific Ara system on a specific machine. See section 5.4.6, "Mobility" for how to exploit this.

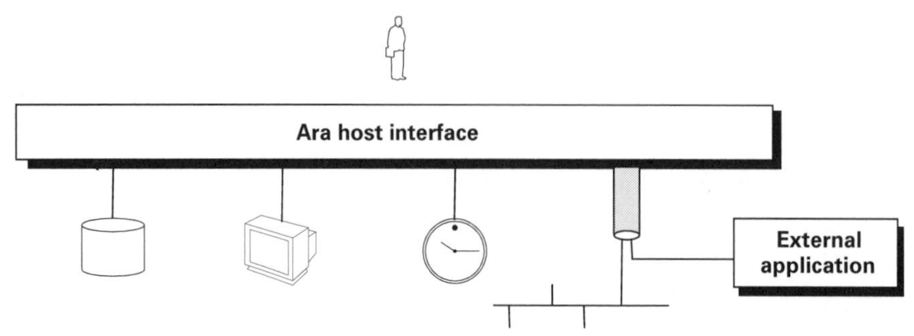

Figure 5.5 Agent access to the host system

Ara agents will have access to a hierarchical file system with flat files, not much different from the interfaces of common file systems. The file system will build on that of the underlying operating system, presenting an enhanced view of it. The main enhancement, besides hiding platform dependencies, will be access control. Specifically, an agent's file access will be authorized, and the volume will be debited to the agent's allowance.[7] Files can be annotated by their owners with access control lists, selectively allowing or denying access to specific agents or classes of agents. Since files are frequently expected to be shared among agents, concurrent-read/exclusive-write locks on files will be provided for the synchronization of concurrent accesses.

In applications such as electronic trade, individual access permissions for specific files are issued or transferred between agents. Since access control lists, being relatively static and modifiable on the local system only, are not always flexible enough for such applications, capabilities for file access are considered as an additional means of authorization. They are more flexible, yet also more expensive to handle.

Capabilities are a kind of voucher for certain files or groups, which can be individually restricted by volume limits or times of validity. Cryptographic certification would make capabilities transferable and tradable across agents and systems.

Access to the graphical user interface (GUI) will be based on Tk, an easy-to-use yet remarkably powerful graphical toolkit closely associated to the Tcl language. Tk will have to be enhanced significantly to make it compatible with Ara features such as concurrency and orthogonal migration within callback functions. To conform with Ara's principle of language independence, it is planned for some later time to separate the Tk toolkit from Tcl, offering it as the general means of graphical user interface access for all agents running on Ara.

A general concept of communication channels will be provided, both for network access and communication with local entities (see the external application interface dis-

---

7 See the introduction of allowances in section 5.2.2.

cussed subsequently). Analogous to the file system, these channels will stay very close to conventional concepts with respect to reading and writing, but will be enhanced by access control features. It is expected that the channel interface will be integrated with the file interface to utilize the access control facilities available there.

Interfaces from client agents to external applications on the local machine will usually be implemented by proxy agents—stationary, trusted agents dedicated to represent the external application to the agent system. A proxy agent supervises the security of all accesses by the client agent and performs potential translations between the two partners. If the client agent is trusted with respect to the external application, and if the application offers a means for external access (such as TCP), the agent may also be allowed to communicate directly through a communication channel without an intermediate proxy.

The Ara core will also offer timing functions going beyond the somewhat inaccurate synchronous logical clock functions currently available, both in the form of synchronous delays and asynchronous timed invocations. The accuracy of these will depend on the timing functions provided by the host system.

At present, since the host interface is not available, agents perform I/O using the standard facilities of the respective language; Tcl agents use Tcl I/O commands, and C agents use the functions of the ANSI-C library.

While being inadequate from the point of view of security, this substitute is sufficient for all tasks which can be handled using the access security mechanisms imposed by the underlying operating system on the Ara system. Note that blocking I/O operations should be used with care, since they currently block the whole operating system process. Both problems will be resolved in the forthcoming Ara host interface implementation.

5.3 Mobile agent applications with Ara

While mobile agents are a fairly general concept for networked computing, there are classes of applications which benefit from their usage. Most prominently, these are applications where a significant amount of data is produced or consumed at a point far away from its source. The prominent example of the remote consumption case is information research: a client searches for information which is scattered throughout the databases of one or more remote servers. Information presentation is the central example for the remote production case, where a server produces presentation information, such as a sophisticated graphical dialog, needed by a remote client. In both cases, rather than transporting the data to its remote destination, an agent can be sent to the data at the remote location, exploiting the ability of mobile agents for computing on site.

Another important area of application is found in mobile computing. The ability of mobile agents to change their host machine during execution can be exploited to let a computation dynamically adapt its location to changing conditions; a mobile agent can enter or leave an area or host shutting down.

These application are now reviewed more closely, illustrating how Ara agents would be used in each.

5.3.1 Information research

An agent wandering through the Internet, looking around, picking up items of interest, occasionally performing a transaction—this is probably the most intuitive and appealing picture that comes to mind when thinking about mobile agents.

Agents are well-equipped for such tasks. An Ara application in need of information at a remote site can send a search agent to meet the server agent managing that information. If the place of the service point is not known beforehand, the search agent can consult the local directory service first.

The search agent then goes to the destination place and meets the server agent at the service point. It proceeds to submit requests to the server, asking for the desired information and receiving replies. The search agent can consult more than one service point, relating and combining the replies.

If the retrieved data indicates that it is worthwhile to extend or move the search to other machines, the search agent makes a note of this and later goes to look there. It could create a clone of itself to access the other site in parallel—arbitrarily sophisticated search strategies can be encoded in the agents. Ara agents benefit from their ability to smoothly migrate and clone when performing tasks distributed across the network.

In many cases, the search agent is actually looking for data more specific than the server interface allows. The agent might ask for a text file, but it is actually interested only in a certain part of it, or a part of the content that satisfies a certain condition. A search agent is well prepared to handle such information filtering by parsing text or relating the information to previous findings. It will later deliver only the desired information, eliminating the transfer of unnecessary data. This is why mobile search agents are especially effective for searching data with complex and irregular structure, such as natural language texts or graphical images: the agent can bring its own data analysis and filtering methods, tailored to and optimized for the specific need.

The value of this approach increases with the structural complexity and the size of the data. In contrast, a remote search is confined to the comparatively primitive filtering methods (if any) offered by the server, such as scanning texts for keywords, and entails the transfer of large amounts of data to the client site. When it comes to concrete programming, the ability to exploit various programming languages proves helpful, employing a Tcl package for text parsing, while using an existing C module for image processing.

At the time of this writing, a realistically sized mobile agent application for researching Usenet news using the Ara system is in development. This agent performs some of the functions that Charles Crizer's scripts presently perform; what Crizer's scripts presently do and what they could do for him in the future are explained in section 1.2.2.

Usenet is a network of autonomous servers, each storing news articles received from its neighbors and passing them on to other neighbors. A user is attached as a client to a local server and may read the server's stock of news articles. Articles which have been on server for a given amount of time are automatically discarded to prevent overflowing the server's storage.

Usenet nodes are autonomous in that each maintains its own stock of articles, propagating and discarding them according to its own policy. Usenet information is thus distributed across the whole network, with each node possessing a specifically local and constantly changing picture of this information. This poses a problem when a user needs to access a set of articles only partially in stock on the local server.

Figure 5.6 illustrates how Ara agents can solve this information research problem. Each Usenet node is equipped with an Ara system providing a server agent for access to the local news server. A user at a node specifies a query for news articles which is used to launch a search agent to collect the articles not present locally. The agent directs its search according to the content of its previous findings.

The agent will track the message's itinerary, visiting servers on that path in the hope of finding further relevant articles which a server has not yet forwarded or not yet discarded. Cross references which link articles will also be exploited by dynamically creating sub-agents to track those recursively.

All search agents will be equipped with a sufficient allowance for their task and will constantly monitor it to prevent getting lost in the network. At the end of their search,

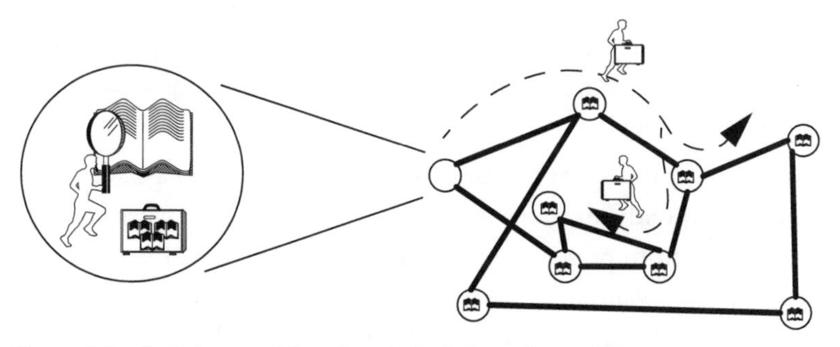

Figure 5.6 Agents searching Usenet for interesting articles

all agents return to their principal, bringing home the collected articles; during the search, the principal does not need to remain online. Without mobile agents, searches of this kind would have to be performed online, and a large number of articles would have to be transferred and processed by the user.

The presented method of searching Usenet is an instance of semantic routing, a concept for distributed information access where a mobile object (a message or an agent) directs its path through one or more intermediate sites according to the content or goal (semantics) of the object. This is a valuable concept in dealing with information which is distributed, providing a flexible and robust method to cope with the constantly changing structure of the global information space. Mobile agents are well-adapted to semantic routing, thanks to their ability to decide autonomously about their further plans and itinerary.

5.3.2 Information presentation

Reversing the perspective, mobile agents can remotely produce information rather than collect it. Agents are beneficial wherever large volumes of data are produced, or when the data depends on interactive user input, since both these scenarios make heavy use of a remote connection which is dispensable with mobile agents. Both requirements of volume and interactivity are clearly combined in remote multimedia presentation: video and audio are displayed to a user in real-time, and user input is reacted upon instantly. Obviously, mobile agents traveling with the data and performing display and input processing at the user's site, are superior to any remote communication solution.

In an application scenario, a server publishing multimedia documents would enclose a custom-written agent to present the data on the client site in an arbitrarily specific fashion. Once a client retrieves and views the document, the presentation agent handles access to the document using the client machine's services and devices in any way it sees fit to present the information. The agent can process the user interaction and take advantage of remote resources behind the scenes. Examples of such presentations include interactive documents, multimedia documents in custom formats such as CAD data, and network-aware documents.

Once again, it is helpful if the agent can be programmed in a language adapted to the specific task. An important advantage of using a customized mobile agent for presentation is the independence from presentation standards, such as HTML, which inevitably lag behind the technical possibilities. Regarding the enormous growth of the Web as a document-publishing medium and the demand for multimedia documents, it is no surprise that using some form of mobile code for document presentation has recently attracted enormous interest. This can be witnessed by the Java language applied in a number of Web browsers and similar tools.

These systems do not presently provide migration; however, presentation agents usually do not move further than their target destination. Mobile code systems without migration are hindered by the lack of this function less noticeably in information presentation than in information research applications.

5.3.3 Mobile computing

Quite apart from specific applications, mobile agents are a useful concept for mobile computing. As a matter of fact, this consideration is what initiated the Ara project. Mobile agents for mobile computers equipped with wireless links are investigated in the project. Actually, the mobile name coincidence here is not a pun. It hints at a real benefit for mobile computers, since some problems caused by hardware mobility can be treated using the software mobility provided by mobile agents.

However, the most prominent benefit of mobile agents for mobile computing is rooted in a better adaption to interaction over wireless links. Sending an agent avoids many of the problems of the physical medium between mobile computers which the traditional model of remote message passing falls a prey to, as sketched in Figure 5.7.

Passing messages between computers effectively creates a remote interaction on a logical connection. This concept, frequently implemented by remote procedure calls, has achieved pervasive use throughout stationary computing networks, and has accordingly been applied to mobile interaction as well. However, the appropriateness of message passing relies on a number of implicit assumptions which become questionable with mobile computers.

First, the presence of an ongoing logical connection between the participants assumes a relatively reliable communication network to keep it up for a sufficient time.

Additionally, such a logical connection presumes that the participants themselves are available for the duration of the interaction. Furthermore, normal communication needs like information filtering and presentation using remote data transfer assumes

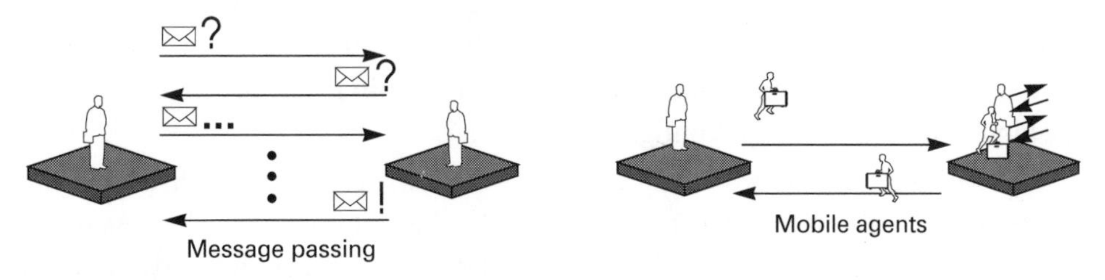

Figure 5.7 Connection usage of remote interaction concepts

reasonable bandwidth. On the whole, remote message interaction assumes a relatively tight coupling between machines.

However, none of the assumptions of reliability, availability, and bandwidth hold for mobile computing systems. Wireless links, especially wide-area ones, are subject to the inevitable disruptions, noise and bandwidth limitations. Mobile devices are also inherently short of energy due to the limited acceptable weight of their batteries; this causes frequent and long-lasting switch-offs, reducing availability. On the whole, mobile units are too loosely coupled to let remote message passing appear as an appropriate basis for mobile computer interaction.

Mobile agents are well prepared for these problems, since they achieve a *decoupling* of the interaction by eliminating the ongoing remote connection. The agents perform interactions coded into their program asynchronously and completely on the destination site, as opposed to the conventional remote and synchronous dialogue. This is particularly welcome for mobile computers, as it requires less bandwidth, is less vulnerable to connection problems and does not require the origin device to be available during the interaction. As a matter of fact, mobile agents take advantage of the available stationary resources on behalf of the mobile unit by dynamically moving a critical portion of the mobile application to the stationary site.

Besides this general reduction of dependence on the connection, up to completely disconnected operation, mobile agents also provide solutions for the problems caused by device mobility itself. The ability to dynamically change their place of execution enables applications like mobile unit substitution and location-specific adaption. When a mobile unit is geographically moving through a wide-area network such as a city, it will encounter diverse organizational domains: office buildings, shopping malls, hospitals, public authority buildings, or company premises.

Each such domain may offer specific local services to guest computers entering the area, ranging from display services like directories, maps, news quotes, or product presentations, to interactive services like navigation and inquiries, up to client components of applications like office automation systems.

Mobile agents can be used to deliver the necessary interface software to entering computers as illustrated in Figure 5.8, automatically adapting them to local customs and opportunities. Such usage of mobile agents highlights their potential to resolve unexpected interface or function mismatches between interacting systems, effectively realizing a kind of dynamic, remote configuration.

Mobile computing networks typically comprise a stationary part of fixed machines connected by wires, and a mobile part using wireless links to mobile devices. As the stationary part provides better availability and connectivity, it is a straightforward idea to shift some functionality from the mobile to the stationary network.

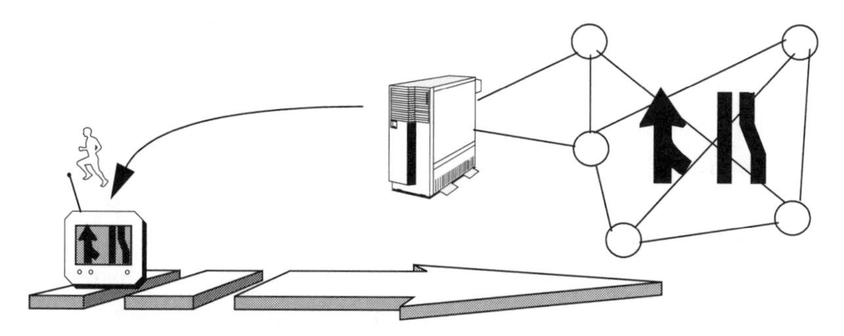

Figure 5.8 Sending a local service interface agent to an entering mobile unit

.Ara agents, being able to move dynamically and smoothly, are particularly suitable for such shifting. Agents staying in the stationary part of the network can act as substitutes of mobile users or applications while their principal is not reachable or prefers not to be contacted, as depicted in Figure 5.9.

A straightforward application for such a substitution is email handling. An agent positioned in the stationary network may screen its prinicipal's incoming email messages, acknowledge receipt, forward selectively, and even reply to certain messages autonomously, not unlike a secretary.

Meanwhile, the principal machine may safely be switched off. The agent can be positioned explicitly by its principal as well as change its location autonomously; it can move when the current host is about to be switched off or requests it to leave. If continuing availability during a transient switch-off period is not required, an agent may also use the checkpointing facility to store itself safely, rather than moving to another machine.

Besides improving availability, a representative agent at the border of the stationary network can also save on transmission costs which can be substantial over wireless links. To this end, the agent can be used to *condense* the data flowing to its principal. Condensation is used here as a rather general notion, to be further defined relative to the concrete application. Transmission costs and delays can be cut further if the agent strives to

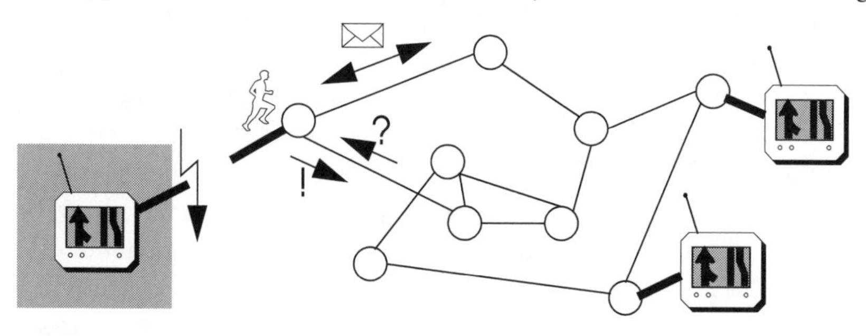

Figure 5.9 A mobile agent acting as the substitute of a mobile unit

stay close to its principal machine, moving along the border of the stationary network while tracking the moving principal. Incidentally, the mobility notions involved in mobile agents and mobile computers actually meet here.

As a demonstration application in mobile computing, a *meeting scheduling* system for mobile users will be realized using Ara, employing mobile agents as substitutes for a human user. The application will offer support for scheduling meetings with other participants using mobile agents which strive to best match the meeting preferences of their human principals and react autonomously to switch-offs and movements of the mobile user. The application will make economic use of wireless communication and handle crashes of participating machines. It will offer a graphical user interface and will be tested in practical operation on mobile computers within the Ara project group.

5.4 Ara programming concepts and features

Now that the basic concepts of Ara have been described and put into perspective, the stage is set to explain how these concepts appear to the mobile agent programmer. This section will describe the Ara core application programming interfaces (APIs) for the two languages currently available for Ara, Tcl and C.[8] The description will be at a more practical level than the other sections (except section 5.5) of this chapter. It will not elaborate on each and every detail; please turn to the manual pages on the CD-ROM for further information.

5.4.1 General conventions of the Ara core interface

Before going right into programming, some preliminaries have to be settled. Ara agents access system functionality by calls to the core API, a set of functions provided by the Ara core for use by agents. Technically, these functions are compiled native code contained in each Ara system. Each is accessible from an agent through a calling interface (called a stub) in the respective language, which is part of the language interpreter (see section 5.6.6). It should be noted here that the overhead of a core call, although similar to an operating system kernel call, is no more than one procedure call (namely the stub). The various stubs for the same core function differ in syntax, but not in semantics. The

8 Note that the MACE interpreter accepts C++ as well, but at the time of this writing, the Ara core offers a C function interface only (a C++ class interface will follow in one of the subsequent releases).

stubs for use by C agents (remember Ara uses a C interpreter for these) are usually named identically to the native core functions such as `Ara_ServicePointCreate()`, since there is a nearly exact correspondence between stubs and core functions. The Tcl stubs, on the other hand, have their own names in part due to the object-oriented calling syntax of Tcl (see "Submitting requests" on page 126 for an example). To distinguish Tcl and C syntax in the following, C function names are prefixed with `Ara_` and suffixed with parentheses like `Ara_Go()`, while Tcl command names are prefixed with `ara_` as in `ara_go`. Compiled agents may use the core functions directly, without stubs.

Note that the term C agent denotes an agent originally written in C and compiled to the MACE bytecode, which now exists in the interpreted bytecode form only. This should not be confounded with compiled agents, which are usually written in C as well, but exist in native machine code form.[9]

The Ara core assigns internal identifications (IDs) to all system objects. Applications reference these objects through these IDs only, in order to protect the objects from undue accesses (see section 5.6.3, "Protection").[10] In the C interface, the type `Ara_Id` is provided for object IDs. This is the return type of all functions returning a system object such as `Ara_ServicePointCreate()`. There is a special value of this type, `ARA_ID_NONE`, designating an invalid ID and used to indicate an error return from such a function. The corresponding Tcl commands throw an error (see below) instead of returning a special value, as is the custom in Tcl.

While it is assumed that the reader has sufficient knowledge of C, one aspect of the Tcl language should be pointed out here: Tcl's only data type is the character string. Accordingly, when types such as `Ara_Id` are mentioned in the following, this refers to the C interface, while the Tcl interface uses a corresponding string representation of the respective type.[11] For example, the C stub of the core function to restore an agent has the signature

```
int Ara_Restore(Ara_Id agent, int mygroup);
```

expecting an object ID and an integer (actually a Boolean value) as a parameter, while returning an integer as the result. The corresponding Tcl stub, on the other hand, expects and returns only strings, interpreted as encodings of the various values.

---

9 Please remember that the MACE interpreter for interpreted C agents is currently not ready for distribution yet (it will be included in the next Ara release). The corresponding stubs, however, are described here nevertheless, both for the sake of completeness and illustration of the native core API.

10 Compiled agents, however, may be exempted here—see section 5.4.9.

11 Please turn to the manual pages of the concerned command for the string representation of a specific type.

Additionally, Tcl allows default parameters, so the Tcl stub corresponding to the C function above has the signature

```
ara_restore ?-mygroup? ?<agent id>?
```
[12]

where <agent_id> is Tcl's string representation of the agent's object ID, and the command's result is the string encoding of an integer. Note that optional arguments, while syntactically required to appear in the C stub, are still called optional with the understanding that special values (usually 0 or ARA_..._NONE) are used when these arguments are meant to be ignored.

A final preliminary remark concerns the error behavior of the API functions. In the C interface, the stubs exhibit the so-called standard error behavior unless stated differently. They flag a potential error by a special return value, usually an integer error code[13] which is ARA_OK on success and ARA_ERROR otherwise. In the Tcl interface, the corresponding stub throws an error and returns an error message as the result. This error message can be retrieved in C as well by means of calling Ara_GetError() after the erroneous call. The error message must be deleted after retrieval by Ara_FreeError(). This might look as follows:

```
if (Ara_Go(...) != ARA_OK) {
    printf("Cannot go ... due to the following error: %s\n",
        Ara_GetError());
Ara_FreeError();
}
```

or, in Tcl, like this:

```
if [catch {ara_go ...} errmsg] {
    puts "Cannot go ... due to the following error: $errmsg"
}
```

Note that the catch command is used in Tcl to catch errors thrown during execution, optionally along with the error message. It is used here to catch any potential errors thrown by ara_go. An uncaught error in Tcl propagates backwards through the calling chain, terminating the current execution if no catch command is encountered.

---

12 ?...? denotes an optional term in Tcl interface specifications, since the more familiar [...] is already used for command result substitution in Tcl.

13 ... or a special "null" value of the return type, e.g. ARA_ID_NONE etc.

5.4.2 *The first steps*

Ara agents are processes, as introduced in section 1.2, "The Life of an Agent." When the Ara system starts up, there is one initial process running, the root process.[14] This is the process which prints the initial prompt to the terminal when the system is run in interactive mode, and which reads and processes the initial input. Interactive mode is the default; when the system is started with a file name argument, it runs in noninteractive (or batch) mode. Input in batch mode comes from the indicated file and the system terminates after reaching the end of the file. In interactive mode, input is read from the terminal until the `ara_shutdown`[15] command is entered. In both modes, the system is also terminated if the root process terminates. For the purpose of this description, interactive mode is assumed, as this is more instructive.

The root process is a Tcl process, which is more convenient for interactive experimentation than C. Any standard Tcl[16] command can be typed to the prompt:

```
% foreach fruit {orange banana} {
+% puts $fruit
+% }
orange
banana
%
```

The user typed a `foreach` command to the `%` prompt, which is a loop over a list of values in Tcl, and the system executed the command, printing out the list elements (as the two lines without a prompt). Note that a + in the prompt indicates that some enclosing command is not yet typed completely.

By the way, there is another version of the prompt which indicates the process currently connected to the terminal. The `ara_toggle_prompt` interactive command is used to alternate between the two prompts:

```
% ara_toggle_prompt
2213942038,1% puts Hello!
Hello!
2213942038,1% ara_toggle_prompt
%
```

---

14 To be precise, there are actually several more processes running, but these are not visible to applications.

15 This command is intended for easy interactive termination only; hence it is not available in batch mode.

16 Tcl version 7.4 at the time of this writing.

Here the figure in the prompt is the printed representation of the current process's object ID. This can be helpful when several processes are connected to the same terminal intermittently.

5.4.3 Basic agent handling

Agent creation

There is one function to create an agent for each agent implementation language. For the sake of uniformity, compiled agents are subsumed here as well; they can be thought of as being implemented in native language. A Tcl agent is created from Tcl by the `ara_agent` stub command, giving it a Tcl script argument to execute, and trailing optional arguments which are passed on to the script as its command line arguments:

```
ara_agent <script> <arg>*
```
[17]

This will create a new agent running in parallel to the other agents in the local system, executing the indicated Tcl script.[18] The newly created agent (in the form of its ID) is returned as the command's result.

The agent command might be used like this:

```
set slave [ ara_agent {
          puts "I'm the slave, my argument is [lindex $argv 0]."
        } \
        foo
      ]
puts "I'm the master, created slave $slave."
```
[19]

This would print something like

```
I'm the master, created slave 2213942038,5.
I'm the slave, my argument is foo.
```

where `2213942038,5` is the string representation of the newly created agent's ID.

Most of the time, agents implemented in a specific language will create only agents of the same language, but agents of different languages may be freely mixed as well. For

17 s* denotes zero or more repetitions of s.

18 The actual command, like many of the following commands and functions, has some additional option arguments which are not described here in order to let the central concepts appear more clearly. The full descriptions can be found in the manual pages.

19 This is a Tcl expression resulting in the value of the first argument of the script.

instance, there is a C stub function equivalent to the ara_agent command, which a C agent might use to create a Tcl agent:

```
Ara_Id  Ara_CreateTclAgent(char* script, int argc, char* argv[],
                           Ara_Allowance global, Ara_Allowance local,
                           int size, flags);
```
[20]

Of course both the Tcl stub and its equivalent C stub map to the same native core function behind the scenes.

Analogous to creating Tcl agents, there is another pair of stubs to create C agents. They are:

```
ara_mace_agent <file name> <arg>*
Ara_Id Ara_CreateMaceAgent(char* fileName, int argc, char* argv[],
                           Ara_Allowance global, Ara_Allowance local,
                           int size);
```

Here the indicated file contains the MACE bytecode the new agent is to execute; again the optional arguments are passed to the agent as command line arguments.

Compiled agents may be created specifying the name of a function *f* which the new agent is to execute. The function is expected to exist in compiled form in an object code file named *f*.o. Some arbitrary parameter data for the new agent can also specified with the stubs:

```
ara_comp_agent <function name> [<arg>]
Ara_Id Ara_CreateCompAgent(char* functionName,
                         byte* data, int dataLength,
                         Ara_Allowance global, Ara_Allowance local,
                         int size);
```

The specified data is a memory area intended to hold parameter data for the new agent. When further languages are added to Ara, these will introduce additional agent creation stubs in the various languages similar to the above.

It is often convenient for an agent to know its own ID. The "me" function returns just this:

```
ara_me
Ara_Id Ara_Me();
```

The agent creation API is completed by the stubs for agent cloning. When cloning, an agent creates a copy of itself, duplicating its internal execution state.[21] Both agents

---

20 The flags and size arguments correspond to optional arguments of the agent command not of interest here; the optional Ara_Allowance arguments will be explained beginning on page 122.

return from the stub; however, the result returned in each case discriminates the two copies. In the original agent, the ID of the new agent is returned, while `ARA_ID_NONE` is the result in the new agent. The signatures reveal that the cloning function is named in reminiscence to the fork system call in the UNIX operating system:

```
ara_fork
Ara_Id Ara_Fork(Ara_Allowance global, Ara_Allowance local,
     int size);
```

Note, however, that this function is used much less frequently in Ara than in UNIX, since the UNIX function is often used to create an actual different process (by subsequent replacement of the new process's memory image), which can be achieved more simply in Ara using an explicit agent creation function.

One final feature of agent creation to be noted here is the treatment of agent allowances. Remember that each agent has a global and a local allowance limiting its resource consumption during its lifetime and at the local place. At the time of creation, the allowances of an agent are determined, and there are two ways to do this; in the first, the new agent is not given a private allowance, but made to share a common allowance with the creating agent. Agents sharing a common allowance form an agent group. A group lives on a common allowance, distributing it among the group members according to their own policy. Shared allowances are the default on agent creation.

In contrast, the second way to determine a new agent's allowance is to give the agent its own private one. This allowance must be specified explicitly on agent creation, and it is deducted from the allowance of the creating agent; the latter agent effectively transfers some of its own allowance (which may come from a private or group allowance) to the new agent.

In this case, the new agent might be viewed as forming a group with only one member. The initial root process is a member of a special group with unlimited allowance. A private allowance can be specified in C by setting the `Ara_Allowance` parameters of the creation stubs to nondefault values. In Tcl this is done by giving optional arguments:

```
ara_agent ?-la <local allowance>? ?-ga <global all.>?
     <script> <arg>*
```

---

21 This implies that the *external* execution state is *not* duplicated (see section 5.2.3, "Agent mobility: going from place to place" for an explanation of this difference). This is a technical restriction explained in section 5.6.5, "Cloning and checkpointing," which may be abolished in a future release. So far, it may be mended by the new agent explicitly recreating the external state, which should be possible to the largest part, because the external environment has not been changed by the cloning procedure.

Note that both the global and local allowances of the new agent must be less than or equal that of the creating agent, and that the local allowance must be less than or equal to the global one. This concept ensures that the total allowance of an agent does not change by distributing it among various subagents or partner agents, so that an agent cannot acquire unauthorized allowance by any means.

As a matter of fact, agent groups are a more fundamental concept in Ara than simply joining agents with a common allowance. Each agent belongs to exactly one group at a time, and the group defines most of the agent's reach and reachability in the system. Groups may form a hierarchy: Whenever an agent is created with a private allowance as described above, the new group becomes a subgroup (or child group) of the creating agent's group.

An agent has a general access right over all agents in its group and in its group's child groups, grandchild groups and so on. In the present Ara implementation, this access right over an agent is not detailed any further, that is, it entitles its holder to any operation whatsoever over this agent; a more fine-grained scheme will follow in a later release.

An agent will leave its group when terminating or moving on to another place; in the latter case, it will find itself the only member of a new isolated group at the new place. When the last member of a group terminates, the group is implicitly deleted and the group's residual allowance for reclaimable resources (such as memory, disk space, system objects) is credited to the allowance of its parent group, if one exists. Note that the current Ara implementation does not provide means to explicitly leave and join a group; this will be available in the next release.

The set of all agents currently existing in the system can be retrieved using the following stubs:

```
ara_agents
int Ara_Agents(Ara_Id** buffer, int* bufferLength);
```

Agent Termination

An agent may terminate voluntarily at any time by exiting. Agents may also make others terminate by killing them, provided the necessary access right. In both cases, an integer value may be left behind as the terminated agent's result:

```
ara_exit ?<value>?
void Ara_Exit(int value);

ara_kill <agent> ?<value>?
int Ara_Kill(Ara_Id agent, int value);
```

A parent agent may retrieve the termination value of another agent by waiting for it; obviously a successful wait also indicates that the other agent has indeed terminated.

Waiting is possible both for a specific agent, giving its ID, and for any agent at all, which is the default.

```
ara_wait ?<agent>?
int Ara_Wait(Ara_Id agent, int* processResultPtr);
```

To end the root process (and this the whole system) in interactive mode, the `ara_shutdown` shortcut can be entered at the prompt.

Agent scheduling

Having populated the system with agents, some functions are needed to control their parallel execution. Ara agents are governed by a simple time-sharing process scheduling model where each agent process is one of three states: *running*, *ready*, or *waiting* (also called blocked). There is always one running process. The others are either ready, meaning they will become running as soon as their they receive a share of execution time, or they are waiting until they are set ready again. The scheduling states are usually managed implicitly by the core, but there are also functions to change an agent's state explicitly, provided the executing agent has access rights over the concerned agent.

In particular, suspending an agent means setting it to the waiting state (if this is the running agent, an immediate switch of control occurs), while activating a waiting agent sets it to the ready state. A running agent may also set itself to the ready state. This is called retiring, effectively performing a voluntary, temporary release of control. The stubs for scheduling read as follows:

```
ara_suspend ?<agent>?  ;# defaults to the running agent
int Ara_Suspend(Ara_Id agent)

ara_activate <agent>
int Ara_Activate(Ara_Id agent);

ara_retire
int Ara_Retire();
```

The core enforces time sharing between the agent processes by preemption,[22] suspending the running process whenever its share is used up and setting one of the ready processes running.[23] It should be noted that once an application engages in explicit suspending and activating, care must be taken to avoid deadlocks; retiring, on the other hand, should be harmless.

---

22 Compiled agents are an exception to this (see section 5.4.9).

23 The current implementation schedules ready processes according to a one-level round-robin policy. Scheduling priorities are expected for a future release.

Note that in the current Ara release, which implements host access by direct operating system calls, agents performing I/O operations (such as waiting for user input from a terminal) retain exclusive control until the operation is completed. Moreover, the time spent in the operation is not debited to the agent's share of execution time. Therefore, agents performing host access operations of unknown duration should retire at regular intervals to preserve parallelism in the system.

An example for this can be found in the system's internal communication process (see section 5.6.2). As another example, interactive agents waiting for terminal input should simply be typed `ara_sleep` at times when the parallel operation of other agents is desired. Besides this temporary requirement for retiring between host access operations, there is another use of this with compiled agents in general (see section 5.4.9, "Compiled agents").

5.4.4 Timing

The timing functions are rather provisional in the current Ara implementation since there is no proper host interface yet—in particular, there is no access to a physical clock, and no facility to react to asynchronous events such as a time interval elapsing. Still, a simple synchronous time service based on a local logical clock is provided, though this is somewhat inaccurate. The current time on the logical clock (a positive integer) may be inquired, and an agent may be suspended for a certain amount of logical time, which is called *sleeping*. Here are the necessary stubs:

```
ara_now
int Ara_TimeNow();
ara_sleep <sleepingTime> ?<agent>?
int Ara_TimeSleep(Ara_Id agent, int sleepingTime);
```

As would be expected, if the agent put to sleep is the current one, a switch of control occurs. However, the obvious question arising here is what unit the logical time is measured in. Lacking physical clock access, the interval of time slice checking (see section 5.6.1, "Processes and internal architecture") is used as a provisional time unit instead of real time. The physical duration of this time unit depends on the concrete platform the system is running on, and its accuracy may depend to some extent on the kind of agents currently running. This will be improved in a future release.

5.4.5 Service points

Announcing and meeting

A service point is created when an agent *announces* it by assigning it a symbolic name. The announcing agent thereby assumes the role of the *server* agent at this service point, and the service point becomes visible under this name for meetings at the local place.[24] An attempt to announce a service point under a name currently in use at the same place will be refused with an error. On success, the newly created service point is returned in the form of its ID:

```
ara_announce <name>
Ara_Id Ara_ServicePointAnnounce(char* name);
```

An agent may *meet* a service point at the local place by specifying its name, thereby becoming a *client* to it. If the desired service point has not been announced yet, the meet operation blocks by default until that happens, but the agent may also use a non-blocking variant which returns an error code in this case. On success, the (ID of the) service point is returned:

```
ara_meet ?-dontwait? <name>
Ara_Id Ara_ServicePointMeet(char* name, int wait);
```

An agent can play the role of a client or server at various service points at the same time, meeting or announcing several service points. A service point can have any number of clients, but only one server, which assumes responsibility for the service point.

Submitting requests

To make use of a service point, a client submits requests to it, to be replied to by the server. The submit operation will return the reply, thus realizing a form of synchronous interaction, implemented by blocking the client until the reply. The syntax and semantics of the requests and replies are up to the client and server: From point of view of the service point, they are simply random arrays of bytes. Requests can be submitted using these stubs:

```
<client's service point> <request>
int Ara_ServicePointSubmit(Ara_Id servicePoint,
                           char* request, size_t requestLength,
                           char** replyPtr, size_t* replyLengthPtr);
```

The syntax of the Tcl stub may seem somewhat peculiar. In fact, this command syntax of naming the object first, followed by the function to be performed on the object, is called object-oriented in Tcl, and is recommended for access to complex objects. The stub returns the reply to the request as its result, provided the server did not

---

24 When the directory service becomes available with Ara, an announcement will optionally make the service point globally visible through this service.

reject it in which case an error is thrown. Note that requests and replies from Tcl agents are necessarily character strings instead of byte arrays, since Tcl cannot represent binary data. Additionally, due to the object-oriented syntax, a few request strings may have actually predefined meaning in Tcl; currently, this is only the `leave` request explained in the section "Renouncing, leaving and closing" on page 128.

The C stub `Ara_ServicePointSubmit()` handles general binary requests and replies, as does the internal service point implementation.[25] The reply is returned in the final two parameters, which may be preset to indicate a memory area intended to receive the reply; this area will be used if its size is sufficient, while a fresh one will be allocated (and returned in these parameters) otherwise. This feature might be used as follows:

```
char replyBuffer[1024];
char* reply = replyBuffer;
size_t replyLength = sizeof replyBuffer;

Ara_ServicePointSubmit(servicePoint, "foo", strlen("foo")+1,
                       &reply, &replyLength);
/* Process data in reply[] */
if (reply != replyBuffer) {
   Ara_Free( reply);
}
```

Fetching and replying to requests

The requests submitted to a service point are queued there until the server *fetches* them. Fetching a request usually blocks until there is at least one to fetch,[26] but there is also an option to return an empty request in that case:

```
<server's service point> fetch ?-dontwait?
int Ara_ServicePointFetch(Ara_Id servicePoint, int wait,
                          Ara_FetchedRequest* fetchedRequestPtr);
```

Again, the `fetch` stub uses the object-oriented syntax. In addition to a reduction of typing, this has the advantage of introducing command name spaces per object type: function specifiers like `fetch` in this example can be reused for other types than service points without name clashes.

A fetched request as returned by the stubs is implemented as a struct object of type `Ara_FetchedRequest` with three components: the request data area, the name[27] of the

---

25 Note that clients should not submit binary data to servers which can handle character strings only and vice versa—the data would be truncated at the first zero byte otherwise.

26 There is also a facility to fetch *all* requests currently pending at a service point, which is not shown here to avoid distraction.

requesting agent, and an identifying token to be used for replying. In Tcl, these components are represented as a 3-element list. The server may use the requestor's name as it sees fit, for example, to decide how much to reply, or whether to reply at all. The following stubs are used to *reply* to a service request:

```
<server's service point> reply <request token> <reply data string>
int Ara_ServicePointReply(Ara_Id servicePoint,
                          char* data, size_t length,
                          Ara_ServiceRequestToken token);
```

If a server decides that it would rather not reply to a specific request, it can *reject* it. This will cause the corresponding client's submit operation to return an error.

```
<server's service point id> reject <request token>
int Ara_ServicePointReject(Ara_Id servicePoint,
                           Ara_ServiceRequestToken token);
```

Renouncing, leaving and closing

A client may end a meeting at a service point by *leaving* it; a server may cancel its announcement by *renouncing* the service point.

```
<client's service point> leave
int Ara_ServicePointLeave(Ara_Id servicePoint);

<server's service point> renounce
int Ara_ServicePointRenounce(Ara_Id servicePoint);
```

Renouncing a service point is the only method to delete it completely from the system. If the service point has submitted requests pending, they are implicitly rejected. When an agent terminates while still acting as a client or server to some service point(s), it implicitly leaves and renounces all these, respectively.

There is one final pair of service point operations: opening and closing. A server may *close* a service point for new requests, with the effect that any request submitted thereafter will immediately throw an error indicating that the service point is closed. Any attempt by new clients to meet the service point will be treated as though the service point had not been announced yet (i.e. these clients are usually blocked). Requests which had been pending or fetched, but not yet replied to, however, remain untouched and may still be fetched or replied to as usual. Closing a service point is intended to

---

27 At the time of this writing, agent names consist solely of the agent's ID, which makes them unique, but conveys no further information beyond identity. Names will be extended in a future release to include structured information about an agent, such as the place and time of creation, identity of its principal, a symbolic description, and so forth.

indicate temporary overload; later, the server may *open* it again, returning to normal operation. Any service point is initially open after announcing it.

```
<server's service point> close
<server's service point> open

int Ara_ServicePointClose(Ara_Id servicePoint);
int Ara_ServicePointOpen(Ara_Id servicePoint);
```

A simple example

This example shows the most common usage pattern of service points, featuring a client and a server agent at a service point. The server agent offers a service point named `Inventory-Service` and expects requests for the currently available supply (as a number) of certain items of interest. The server serves such requests until requested for a pseudoitem named `finished`. The client agent in the example meets this service point, requests the supply of items named `small-objects`, and leaves. Here is the example Tcl code for the server and the client:

```
set supply(small-objects)  20
set supply(medium-objects)  8
set supply(large-objects)   3

set sp [announce Inventory-Service]
puts "This is the Inventory server, waiting for requests..."
set request [$sp fetch]; # Request format: {data token client-name}
set item [lindex $request 0]
while {[string compare $item finished]} {
   $sp reply [lindex $request 1] $supply($item)
   set request [$sp fetch]
   set item [lindex $request 0]

$sp renounce
puts "Inventory service closed down."


set sp [meet Inventory-Service]
if {$sp == ""} {
   puts "Cannot meet Inventory Service - exiting."

   exit
set result [$sp medium-objects]
puts "This is the client, found $result medium objects in supply."
$sp leave
```

Running the server and the client as agents at the same place will produce the following output:

```
This is the Inventory server, waiting for requests...
```

```
This is the client, found 8 medium objects in supply.
```

The reader may guess what would happen if the client executed `$sp finished` before leaving. To illustrate the correspondence between the Tcl and C interfaces, here is the same server coded in C:

```
static int GetSupply(char* item)
{    /* Return the number corresponding to item */  }
...
Ara_FetchedRequest request;
char replyString[20];
Ara_Id sp = Ara_ServicePointAnnounce("Inventory-Service");
puts("This is the Inventory service, waiting for requests...");
Ara_ServicePointFetch(sp, 0, &request);
while (strcmp(request.data, "finished")) {
    sprintf(replyString, "%d", GetSupply(request.data));
    Ara_ServicePointReply(sp, replyString, strlen(replyString)+1,
                          request.token);
    Ara_ServicePointFetch( sp, 0, &request);

Ara_ServicePointRenounce( sp);
puts("Inventory service closed down.");
```

5.4.6 Mobility

The `go` operation for migration has been introduced before as the means for an agent to move between places. While the basic act of moving is as simple as it seems, some technical issues still have to be explained in order to utilize all its functions. The full interfaces read as follows:

```
ara_go ?-la <local allowance at destination>? \
       ?-ga <global allowance at destination>? \
       <destination place name> ?<agent>?

int Ara_Go( Ara_Id agent, Ara_PlaceName destination,
            Ara_Allowance local, Ara_Allowance global);
```

An agent is always staying at a place, and the migration operation will make it move to another place as named[28] in the destination argument. Agent mobility can also be exercised on a single machine by making agents move between places located on the same machine. Remember, however, that the current Ara implementation provides only

28 The API for place names will be explained in the subsequent section 5.4.7.

one place per system; as a temporary fix, this can be circumvented by starting several separate Ara systems on one machine,[29] each system with its own place. Agents can then move between the places of these systems within the same machine.

When an agent goes to another place, it must take an allowance for its expenses there. If the agent has a private allowance, that is, it does not share it with a group, it takes this along completely by default. If, however, the agent shares its allowance with a group, it should indicate on moving how much of the group allowance it wishes to take along. This is the purpose of the global allowance argument of the migration operation; it defaults to the group allowance divided by the number of group members. Note that this default results in the complete allowance in the common case of a singleton group, which is what is usually desired. The agent leaves its group on migration, and the indicated allowance is deducted from that of the group to become the agent's private one.

Remember that agents can also have local allowances valid for their current place only. Therefore, besides taking along a global allowance as described, the agent may also specify a local allowance on migration, to be valid at the destination place (default is the full global allowance). At the time of this writing, the local allowance specified by an arriving agent is always accepted by the receiving place (since there is currently only one default place per system and its policy is to accept all agents and allowances).

There are two restrictions to the migration operation. First, as mentioned before, the external state of an agent (its relations to other system objects and resources like service points, files, or windows) are not migrated along with the agent, since those objects depend on the local machine. In particular, a migrating agent loses its relation to its group. Second, there is a somewhat obscure restriction on migration of agents programmed in Tcl: they are not allowed to migrate from within Tcl's traces and asynchronous `proc`'s. This is due to technical reasons and might be abolished in the future, but should not pose a real problem.

Finally, the agent argument to the migration operation remains to be explained. This can be used to specify *which* agent is to go. Of course this defaults to the running agent, but the migration mechanism in Ara has been implemented to also allow migrating another agent, asynchronously to its execution, provided the active agent possesses the access right over the migrated one. This feature may be used as a building block for emergency measures or load distribution. Note however, that the concerned agents must be prepared to such migrations in some form.

---

29 Note that the various Ara systems on the same machine have to be started each with its own environment settings to prevent collisions. This is explained in the installation guide on the CD-ROM.

5.4.7 Place names

Place names are not system objects, but application data, in order to allow agents taking them along on migration. However, their exact format is hidden from the application by a small interface, which is expected to grow as place names become more structured in future system releases.

In C, place names are represented as objects of type `Ara_PlaceName`. Objects of this type will usually be obtained from the directory service once it is available, but they also have a character string representation, which can be used by applications to construct place names.[30] Conversion between `Ara_PlaceName` and a string representation is performed by these functions:

```
Ara_PlaceName Ara_PlaceNameCreate(char* initValue);
void  Ara_PlaceNamePrint(Ara_PlaceName placeName,
     char* outputString);
```

In Tcl, which is string based, the string representations must be used directly, and conversions are performed internally. Note that in C an `Ara_PlaceName` which has been constructed should be deleted using `Ara_PlaceNameDelete` to avoid a memory leak.

An agent can find the name of its current (local) place through the following stubs:

```
ara_here
Ara_PlaceName Ara_Here();
```

Place names can be compared for identity, returning a Boolean result, by:

```
ara_placename equal <place name> <place name>
int Ara_PlaceNameEqual(Ara_PlaceName, Ara_PlaceName);
```

The C interface also provides functions to delete, copy, and assign place names; again, this makes no sense for Tcl. Note that place names are dynamically created objects; they must be finally deleted using `Ara_PlaceNameDelete()` to avoid a memory leak.

The stubs look as follows:

```
void Ara_PlaceNameDelete(Ara_PlaceName);
Ara_PlaceName Ara_PlaceNameCopy(Ara_PlaceName source);
void Ara_PlaceNameAssign(Ara_PlaceName* dest, Ara_PlaceName
      source);
```

30 In the current Ara implementation, providing only one place per system and basically only TCP as a communication protocol, the string representation of place names is *machine*[:*port*], *machine* being the DNS name or IP address of the destination machine, and *port* being the TCP port of the target Ara system's default place. section 5.7 gives a brief outlook on the expected future format of place names.

5.4.8 Checkpointing

An agent may create a checkpoint of its current (internal) state using one of the following stubs:

```
ara_checkpoint ?-exit ?<exitVal>?? ?-ga <global allow.
        after restore>? ?-la <local...>?
int Ara_Checkpoint(int* restored, int exit, int exitValue,
                    Ara_Allowance global, Ara_Allowance local);
```

Similar to the forking function of "Agent creation" on page 120, this function returns in two different contexts: after creating the checkpoint, control returns like any normal call. If the agent is ever restored from that checkpoint later, it will resume control returning from this function as well, indicating the fact of restoration by a different result: the Tcl stub returns `1` after a restoration, and `0` on normal return; the C stub returns the same in the `restored` output variable, and shows standard error behavior for the rest.

The checkpoint is stored as a disk file in a dedicated directory under a name derived from the ID of the checkpointed agent. This file is kept until the checkpoint is discarded. The allowance parameters define the allowance for the agent after a potential restoration; the defaults are the same as with migration. Note that this allowance is stored with the checkpoint, and deducted from the allowance of the checkpointing agent.

Finally, it is a common case that an agent plans to exit immediately after checkpointing itself; the exit flag provides just this, optionally leaving a termination value as usual. This flag is not only a convenience, but the only way for an agent to store its *complete* allowance in the checkpoint, since any execution continued after checkpoint creation, no matter how short, would require its own share of the available allowance.

Restoration works by specifying the ID of the agent to be restored; the corresponding file is then used to re-create the checkpointed agent, and the checkpoint file is deleted. The restored agent is placed in its own isolated group by default, but it may also be received into the restoring agent's group. The agent to be restored defaults to the current one. Note that restoring the current agent implies the termination of its present incarnation, since that is replaced with the incarnation from the time of checkpointing. The stubs read as follows:

```
ara_restore ?-mygroup? ?<agent>?
int Ara_Restore(Ara_Id agent, int mygroup);
```

When restoring another agent, the Tcl stub returns nothing, while the C stub shows standard error behavior. Obviously, in the case of restoring the current agent, restoration does not return. Remember that restored agents cannot expect to find their external state unchanged (this was explained in section 5.2.2). When restoring another agent, the application should either make sure that no incarnation of this agent has survived anywhere, or be prepared to resolve such collisions.

When checkpoints are no longer needed for potential restorations, they can be discarded as follows, deleting the corresponding file:

```
ara_discard ?<agent>?
int Ara_Discard(Ara_Id id);
```

When discarding a checkpoint, the allowance stored with it is recycled by crediting it to the discarding agent's allowance. Discarding returns a Boolean result indicating whether such a checkpoint had indeed existed.

Restoring and discarding another agent's checkpoint do not perform any access right checking in the current implementation.

The most common use of checkpointing is to create a fall-back line on which to recover after a fatal failure. An agent might use this before going to a dangerous place, registering at a wake-up service to restore it unless the registration is canceled within a certain time:

```
if {![ara_checkpoint]} {
    $wake-up register [ara_me] 100 ;# "Wake me up after 100
      time units"
    set back [ara_here] ;# The name of the local place
    ara_go $away
    ... ;# Do some dangerous work at $away
    ara_go $back
    $wake-up cancel [ara_me]
} else {
    # Oops, did not return in time from $away - start recovery,
    # e.g. find out if the agent has crashed.
    ...
}
```

Another common usage is to guard dangerous procedures with checkpoints, under the assumption that some other agent will restore the failed one:

```
proc p_guarded {...} {
    if {[ara_checkpoint]} {
        return "Failed in proc p!"
    } else {
        p ;# Do the real work
        ara_discard
} }
```

5.4.9 Compiled agents

Compiled agents provide a native speed alternative to the standard interpreted Ara agents for cases where security and portability requirements are not strictly necessary. Most prominently, this applies to trusted agents which are resident at a site. This is the way to extend an Ara system with specific resident functionality such as a new service, to be offered to visiting agents. The service is coded as a server agent (in C, presumably) and compiled to native code, yielding a compiled agent. This can be performed at run-time when an agent requests that service.[31] While the core is common to all Ara systems, the set of stationary server agents is what distinguishes individual sites, potentially ranging from hand-held devices with only a few server agents to corporate installations with thousands of them.

Compiled agents are different in two respects, apart from their increased execution speed. They cannot normally migrate, and the security of their execution cannot totally be warranted. Regarding common usage as described above, these restrictions do not pose a problem. Apart from migration and security, they can be treated like any other agent: the complete API as introduced in section 5.4 applies to compiled agents as well. The lack of migration capability is rooted in the very nature of native code, operating on a specific microprocessor, which precludes a direct transfer to a machine with a heterogeneous architecture. Actually, migration of compiled agents has been implemented in an indirect way, by exploiting the source code, provided it is available.[32] The source code is transparently moved along with the migrating agent and recompiled at each destination.

Compiled agents are not executed within an interpreter, but have access to the physical processor, memory, and operating system. They might write to arbitrary memory locations and perform unpredictable system calls. It is difficult to enforce (efficient) run-time checks upon them while staying independent of the operating system and hardware. For this reason, Ara does not currently perform any run-time checking of compiled agents. It is, therefore, recommended to load compiled agents obtained from trusted sources. In the case of a remote source, this means a digitally signed transfer[33] from a trusted site. For compiled agents loaded from the local file system, it must be ensured that the agent is from a location protected against access by untrusted agents.

However, the system does offer some protection against undue access by compiled agents using a creation-time check of the external functions referenced by the agent: The place where the agent is to be created checks the set of external functions referenced in

---

31 Compiled agent creation is implemented using dynamic link editing of the running Ara system.

32 The source code is assumed to be contained in a file named *f*.c, where *f* is the name specified at agent creation (see "Agent creation" on page 120).

33 Support for this will be added in a future Ara release.

the agent code against a specific set of functions it deems allowed for this agent. If the agent is discovered to reference a disallowed function, the creation is refused. Note that this is more of a protection against misunderstandings than a bullet-proof guarantee, since a malicious compiled agent might also covertly call functions without explicitly naming them, using direct calls to their entry addresses.

Each place can employ many such sets of allowed functions, corresponding to specific applications or agent sources with varying degrees of trust. Note, however, that the system default place does not perform allowed function checking (yet).

Besides the risk of undue access, another security concern with compiled agents is their possession of control once the system enters compiled code. A compiled Ara agent retains control until voluntarily and synchronously releasing it. This implementation was chosen since an asynchronous, forced interruption would leave the agent in a machine-dependent state, complicating its further handling, while gaining nothing for interpreted agents. For this reason, compiled agents must be trusted to return control to the system frequently enough to preserve parallelism among the agents in the same system. This can be done by retiring at regular intervals.

Finally, the API for compiled agents is different in three syntactical or technical respects from that used by C agents. First, once they are started, compiled agents are trusted as explained previously. This implies that the security measure of protecting system objects by an indirection through object IDs no longer makes sense. Accordingly, compiled agents usually access system objects directly through memory addresses (i.e., C pointers). Instead of having one `Ara_Id` type for all objects, there are separate types such as `Ara_ServicePoint` which are returned by object creation functions and expected by object access functions. Special values such as `ARA_SERVICEPOINT_NONE`, are used in place of `ARA_ID_NONE` to designate null values.

Second, within the Ara core, agents are usually called processes for historical reasons. This is mirrored in the naming used in the compiled agent API, for example, the object type for agents is called `Ara_Process`. This difference bears no further significance: whenever process is used, as in `Ara_Kill(Ara_Process victim)`, it is safe to simply think of an agent.

Third, the core function to create compiled agents has a slightly wider interface than the C stub given in the section "Agent creation" on page 120, allowing the creation of processes executing code which already exists in memory in compiled form (instead of loading it from a file):

```
Ara_Process Ara_CreateCompProcess( Ara_CompCode code, char*
 functionName,
                                   byte* data, int dataLength,
                                   Ara_Allowance global, Ara_Allowance
                                   local, int size, int flags);
```

The code parameter may be used to specify the function to be executed by the new process; functionName must then be NULL. Ara_CompCode is a type definition for int (*)(byte*, int).

5.4.10 *Allowances*

Allowances have played a role several times already as arguments to the agent creation and migration operations, the details of which have been covered in their respective sections. Accordingly, the purpose of allowances to limit an agent's resource consumption, and their treatment in agent creation and migration, as well as the concept of global and local allowances should be clear by now. What remains to be described is the API for explicit access to allowances. It should be remarked here that allowances, including groups, are a recent concept in Ara, and not completely stable.

Remember that at the time of this writing, allowances are defined for execution time and memory consumption only; other resources like disk space, system objects created (i.e. agents, service points, network connections, and so forth.), or visited places will follow. An agent can *inquire* about its own current local and global allowance at any time. An agent can also inquire about other agents' local and global allowances, providing that the agent making the inquiry has access rights to the other agents' information. This might be used to choose between actions of differing resource requirements, or to check whether the local place has indeed honored the local allowance requested on entering it. The stubs look as follows:

```
ara_get_allowance ?<agent>?
int Ara_GetAllowance(Ara_Allowance* global, Ara_Allowance* local,
                     Ara_Id agent);
```

Ara_Allowance is a C struct containing the individual allowances for execution time and memory as members, while the Tcl representation of allowances is a 2-element list of these numbers. The ara_get_allowance stub returns the global and local allowance as a list of two such allowances.

Not surprisingly, it is not possible to arbitrarily set an allowance, a security measure to prevent tampering by greedy agents. There is, however, an operation to *transfer* some allowance from one agent to another, providing the necessary access right. If the two agents share a common allowance a transfer has no effect.

The transferred amount is deducted from the source agent's allowance, and added to the receiving agent's allowance. This ensures that allowances can be distributed among agents, yet the total sum of all allowances cannot be increased. Note that the amount transferred can be positive or negative, corresponding to giving and taking.

Allowance transfers are mostly intended for cooperating agents (re-)distributing resources among each other, and agents running out of resources having fresh allowance transferred to them from their home site. Other uses of allowances are conceivable, such as objects of trade between agents. The offer for a remote service might be bundled together with an allowance sufficient to cover the travel and carry out the access, to make the offering independent of the expenses involved with its use.

When transferring allowances between two agents, it is important to discriminate between transfers with respect to the local, global, or both variants of the allowance. As a general prerequisite, the transferred amount is deducted from the source agent's allowance, so the latter must be sufficient to cover the transfer. This assured, transfers of purely global allowances are performed without further questions. Transfers of local allowances also require that both agents are located at the same place, in order to preserve the total local allowances issued by this place; further, the amount transferred must stay within the bounds of the recipient's global allowance. Note that a transfer of purely local allowance does not change the global allowances of either source or recipient agent; a simultaneous transfer of both variants may be used for this. Here are the stubs for allowance transfer; the source agent defaults to the current one:

```
ara_transfer_allowance ?-la <local allowance>?  ?-ga <glob.
      allowance>?\
               <recipient agent> ?<source agent>?
int Ara_TransferAllowance(Ara_Allowance global, Ara_Allowance
      local,
                 Ara_Id sourceAgent, Ara_Id recipientAgent);
```

As mentioned before in the discussion of agent creation and migration operations, agents are implicitly assigned a group at creation time, possibly sharing a common allowance. At the time of this writing, there are no operations to explicitly leave[34] or enter a group. Presently, an agent can find the size, (the number of members of its own) group as follows:

```
ara_groupsize ?<agent>?
int Ara_GroupSize(Ara_Id agent);
```

At the time of this writing, there is no facility for agents to catch allowance exhaustion; instead, agents having exhausted their allowance will be terminated by the system. This should not be a severe problem, since an agent can, of course, inquire about its current allowance at any time and direct its actions according to that. In a future release, agents will be able to register arbitrary code to be executed asynchronously when

34 Remember that a migrating agent implicitly leaves its group behind.

resources run low, to take appropriate measures to avoid being terminated such as leaving the system or having additional allowance transferred to it.

5.4.11 *Dynamic memory*

Languages with explicit memory management need a facility to dynamically obtain raw memory from the system. C is an example for such a language, while Tcl is not. As memory consumption must be accounted to the consuming agent's allowance, all dynamic memory allocations must be performed through dedicated core functions. The C API provides stubs for allocation and deletion of a memory block, the two classical memory management functions:

```
void* Ara_Alloc(size_t size);
void Ara_Free(void* p);
```

As usual, memory blocks allocated by `Ara_Alloc()` must be freed using `Ara_Free()` in order to reuse that memory. Unfreed memory will be freed implicitly on agent termination. Note that while languages without explicit memory management need not use these stubs, their memory consumption is accounted for nevertheless, since in that case the interpreter uses the concerned core functions implicitly.

5.4.12 *Input and output*

As explained in section 5.2.4, "Accessing the host system" Ara does not yet provide I/O functions of its own, and allows applications to use their own native I/O facilities instead. For instance, a Tcl agent may read from a file as usual by:

```
set file [open myfile]

read $file

close $file
```

while a C agent would use the familiar

```
FILE* fp = fopen("myfile", "r");
fread(buffer, size, n, fp);
flose(fp);
```

5.5 A programming example

Having looked at the Ara programming features in some detail, it is helpful to see the complete picture as an actual application with full source code.[35] In the limited space available, this section will present a small demonstration application featuring a mobile agent to search the World Wide Web for interesting documents. The agent will visit sites, examine their data, collect results, and continue its itinerary according to its findings. In terms of section 2.1, this is a typical information research application.

The strategy of the search agent is to move along the hyperlinks found in interesting documents, in the hope that these might lead to other interesting documents, as is often the case given the roughly content-based topology of the Web. The search continues until a predetermined allowance of resources has been consumed, or until there are no more interesting links to be examined. As the result of its search, the agent brings a list of the URLs of all discovered documents back to its home place. To provide access to the documents of a site, a stationary server agent is employed as a document retrieval service; when presented with the URL of a document, it will reply with the contents of this document.

Conceptually, this server is a functionally restricted proxy agent (in terms of section 5.2.4) of the local Web server daemon; however, to keep the example short, the presented server agent accesses the documents directly through the file system, rather than through conversation with a Web server.

The question remains of how to define what exactly makes a document interesting in terms of this search. Incidentally, the agent is to search for documents about mobile agent technology. Many sophisticated conditions are conceivable to delimit the desired content, but rather than digressing into text analysis here, it shall suffice for the sake of illustration that the document contain the string mobile agent.

5.5.1 The document server agent

Being stationary, the server is implemented as a compiled agent for improved performance; the implementation shown here is written in C. The CD-ROM also contains an equivalent implementation in Tcl for the sake of comparison. Basically, the server announces a service point named `Document Retrieval Service`, and then fetches requests for documents in a loop until termination. Each request is expected to have the format of a path name of a document file in the server's local file system, and the

35 The source code is also contained on the CD-ROM.

contents of the corresponding file are read and returned to the client as a reply. This results in the following overall structure:

```
serviceP = Ara_ServicePointAnnounce("Document Retrieval Service");
do {
      Ara_ServicePointFetch( serviceP, ARA_SERVICEPOINT_WAIT, &request));
      normalizedPathName = Parse document file name from request.data;
      length = size of document file;
      replyBuffer = Ara_Alloc(length);
      fread( from file named normalizedPathName, into replyBuffer);
      Ara_ServicePointReply( serviceP, replyBuffer, length, request.token);
      Ara_Free(replyBuffer);
   } while not finished;
   Ara_ServicePointRenounce( serviceP);
```

The full source code is now presented in detail. The server is implemented as a main function `Server_Main`, which serves as the top-level function *f* of the underlying compiled agent upon agent creation (see "Agent creation" on page 120). As is common in C, the source code begins with some inclusions of needed interfaces; note in particular the inclusion of `ara.h`, the Ara core API for compiled agents. The main function commences with the definition of three macros required by Ara for handling the function's state (see section 5.6.4 and the documentation on the CD-ROM), which are not of further interest here, as they are defined to be empty.

```
#include <stdio.h>
#include <string.h>
#include <ara.h>

int Server_Main (data, dataLength)
  byte* data;
  int dataLength;
{
# define Ara_ParcelLocalState(process)
# define Ara_UnparcelLocalState(process)
# define Ara_CleanupLocalState
```

After defining the needed local variables, another Ara state handling macro follows which marks the beginning of the function's code. The server initializes itself by determining the path prefix of the documents it wishes to provide for client access. All such documents are assumed to reside in the file system subtree rooted at this prefix, and any document file name requested by clients will be interpreted as relative to the prefix. Incidentally, the prefix in this example is chosen as the directory the server was started from, to allow for easy experimentation with different prefixes. The initialization is completed by announcing the service point for delivering the documents, whereupon the server enters the loop to wait for document requests.

```
# define EMERGENCY_STRING "Out of memory"
  char prefix[1024];
  char* normalizedPathName;
  Ara_ServicePoint servicePoint;
  Ara_FetchedRequest request;
  char requestBuffer[1024];
  char* replyBuffer;
  FILE* stream;
  int length = 0;

  Ara_DeclAndCheckSwitch1;

  getcwd(prefix,sizeof prefix);
  strcat(prefix,"/");
  server = Ara_ServicePointAnnounce("Document Retrieval Service");

  do {
```

Each pass through the loop begins with fetching a request from the service point; the server uses the blocking variant of fetching in order to wait until at least one request is available. The macro enclosing the core call for fetching is part of the state handling again. Requests are fetched using an `Ara_FetchedRequest` struct object, which also refers to a memory area to receive the request data. This area may optionally be initialized with one prepared by the server for this purpose, as is shown here with `requestBuffer`. In any case, the `Ara_FetchedRequest` will contain the request data on return from fetching, and the prepared memory is used, provided it is large enough. After fetching, a memory buffer to hold the physical path name of the requested document is allocated.

```
request.data = requestBuffer;
request.length = sizeof requestBuffer;
Ara_SwitchCall1(Ara_ServicePointFetch(servicePoint,
                          ARA_SERVICEPOINT_WAIT, &request));
normalizedPathName = Ara_Alloc(request.length+strlen(prefix)+1);
```

Now the path name buffer is filled with the requested file name, appended to the subtree prefix described above. If the request data space had indeed been freshly allocated during fetching, it may be deleted now. The path name is subjected to some parsing, and it is checked that it still bears the prefix after that; this is a security measure explained with the parsing function below. Now the requested document file can be opened, and it is read in whole into a freshly allocated buffer. A trailing zero byte is appended to make the data easily handled as a character string in the case of a text file:

```
if (normalizedPathName != NULL) {
    sprintf(normalizedPathName,"%s%s", prefix, request.data);
    if (request.data != requestBuffer) {
            Ara_Free(request.data);
   }
   if (   ParseFileName(normalizedPathName) != NULL
      && !strncmp(normalizedPathName, prefix, strlen(prefix))) {
      if ( (stream = fopen(normalizedPathName, "r")) != NULL) {
       fseek(stream,0L,2);
       length = ftell(stream);
       fseek(stream,0L,0);
       if (length < strlen(normalizedPathName)+80) {
         replyBuffer = Ara_Alloc(strlen(normalizedPathName)+81);
       } else {
         replyBuffer = Ara_Alloc(length+1);
       }
       if( replyBuffer == NULL) {
         replyBuffer = EMERGENCY_STRING;
       } else if (!fread(replyBuffer,1,length,stream)) {
              sprintf(replyBuffer,"Error while reading file %s",
                                normalizedPathName);
           }
    replyBuffer[length] = '\0';
    length += 1;
    fclose(stream);
```

If all went well, the reply data is now assembled in the `replyBuffer`. If errors occurred in accessing the file, an error message is stored in the `replyBuffer` instead as follows:

```
} else {
    replyBuffer = Ara_Alloc(strlen(normalizedPathName)+80);
    if( replyBuffer == NULL) {
          replyBuffer = EMERGENCY_STRING;
    } else {
          sprintf( replyBuffer,
            "Illegal operation: File %s does not exist",
            normalizedPathName);
    }

     length = strlen(replyBuffer);
  }
 } else {
   replyBuffer = Ara_Alloc(strlen(normalizedPathName)+80);
   if( replyBuffer == NULL) {
     replyBuffer = EMERGENCY_STRING;
   } else {
    sprintf(replyBuffer, "Illegal operation: Permission denied");
```

```
      }
      length = strlen(replyBuffer);
   }
   else {
      replyBuffer = EMERGENCY_STRING;
      length = strlen(replyBuffer);
   }
```

The processing of a request is completed by passing the retrieved data as a reply to the service point, and freeing all buffers:

```
Ara_ServicePointReply(server, replyBuffer, length,
         request.token);
Ara_Free(normalizedPathName;
if( !strcmp(replyBuffer, EMERGENCY_STRING))  {
   Ara_Free(replyBuffer);
}
```

The server will now begin the next pass through the loop, until it terminates. In this example, the server agent can be terminated by submitting a special request `exit` to the service point. This is only for convenient experimentation; a real server would presumably not let itself be terminated by a client, but would rather decide this on its own based on additional conditions, or be terminated by some superior controlling agent. Upon termination, the service point is renounced, and the main function ends with some macro undefinitions corresponding to those at the beginning.

```
} while(strcmp(request.data, "exit"));

Ara_ServicePointRenounce(servicePoint);
return 0;

# undef Ara_DeleteLocalState
# undef Ara_UnparcelLocalState
# undef Ara_ParcelLocalState
}
```

The last item to report of the server is the auxiliary function used for file name parsing. This function removes any pseudodirectories named `..` from a path name by "performing" their effect on the rest of the path. This is a security measure on the server's side, since only files in the file system subtree designated for client access are to be exposed. A malicious client could try to subvert this restriction by inserting `..` into the file name, effectively reaching out of the designated subtree. This is the source code of the function:

```
static char* ParseFileName( pathName)
  char* pathName;
{
```

```
    char* help;
    int i, j;
    while (help = strstr( pathName, "//")) != NULL) {
           j = help - pathName-1;
           while( j >= 0 && pathName[j] != '/') {
               j --;
  }
           if( j == -1) {  Ara_Free( pathName);
             return NULL;
  }
   for(i=help-pathName+3;  i<=strlen( pathName); i ++, j++) {
             pathName[j] = pathName[i];
           }
  }
  return pathName;
  }
```

5.5.2 The search agent

The search agent is implemented in Tcl, which is a language well adapted to text processing. Roughly, the agent works through a list of URL references to documents to be searched, moving to each document's site, and retrieving its content from the local document retrieval service (see former section). Each retrieved document, provided it turns out to be interesting as described above, is searched then for hyperlinks referencing further documents, and all such references are added to the work list for later inspection. Additionally, all references to interesting documents are collected, and this collection is printed as the search result upon returning to the home site. At regular intervals, the residual allowance is checked if it is still sufficient to continue the search.

The overall structure of the agent thus looks like this:

```
    Initialize and ask the user for an initial list of urlsToVisit to start the search at
SearchWeb:
  while (there are urlsToVisit and AllowanceSufficient) {
    go  site of next document
    set servicePoint [meet "Document Retrieval Service"]
    while  (there are urlsToVisit  at this site and AllowanceSufficient) {
      set document  [$servicePoint  path of next URL]
      if  ($document contains "mobile agent") {
        Collect this URL
        SearchDocument  $document, extracting URLs and adding them to urlsToVisit
      }
    }
    $servicePoint leave
  }
go $home
PrintResult
```

The full source code of the agent's main program begins with the initialization of the global variables, the purpose of which is explained in the source code comments. Most important is `urlsToVisit`, the general list of references still to be searched. At the beginning of the program, this list is filled with some document URLs typed in by the user; simple site names, which are interpreted as references to the top-level document at that site may also be typed here.

```
set filesRead ""    ;# contains all files already read
set document ""    ;# the content of the currentFile
set urlsToVisit "";# list of files (site/path) still to read
set currentSite "";# name/address of the Site currently worked at
set currentFile "";# file (incl. path) currently examined
set filesHere ""  ;# files found on this site which are still to be
                  # examined; files contain the path without site
set home [ara_here];# the site to return to after work
                 (the start site)
set sufficientTime 0 ;# flag indicating after returning home whether
                     # execution time was sufficient
set sufficientMemory 0 ;# the same for memory

puts "Please enter the machine(s) or URL(s) to start the search at:"
while {[gets stdin next]} {
  if {[regexp -nocase {http://} $next]} {
    # Assume that $next is a URL to start the search at.
    regexp {http://(.*)} $next dummy next
  } else {
    # Assume that $next is a machine to be searched; start at
                 its index
    # document.
    set next "$next/index.html"
  }
  lappend urlsToVisit $next
```

After initialization, the Web search is started. The agent is instructed to search for documents matching the (most simple) regular expression `mobile agent`, and to finish the search early if its residual allowance of execution time ever falls below 300 time units, or if its memory allowance falls below 100KB. After returning home, the results are printed, and the user is notified if the agent had to finish early, as opposed to completely covering all reachable documents:

```
SearchWeb {300 100k} "mobile agent"
ara_go $home
PrintResult
if { !$sufficientMemory} {
  puts "Gone out of memory allowance"
  puts "Total allowance left: [ara_get_allowance]"
```

```
}
if { !$sufficientTime} {
  puts "Gone out of time allowance"
  puts "Total allowance left: [ara_get_allowance]"
}
exit
```

Since the agent will return when its allowance runs low, it should be started with sufficient allowance in the first place, the exact amount of which will, of course, depend on the size of the search area. The example code on the CD-ROM comes with a test Web and a corresponding initial allowance.

The heart of the search agent is the `SearchWeb` procedure, basically consisting of two nested loops. The outer loop processes the general list of URLs still to be examined, usually located at various remote sites. Once moved to a specific site, the agent meets the local document retrieval service, and retrieves and searches all known documents located at this site. Note also that before each loop entry, the agent checks its residual local allowance to leave in time before this runs low. In this example, the agent returns home due to a low allowance:

```
proc SearchWeb {minAllowanceNeeded} {
   global document
   global filesHere
   global filesRead
   global currentSite
   global currentFile
   global urlsToVisit
   global sufficientTime
   global sufficientMemory

   while {   ([llength $urlsToVisit] > 0)
         && [AllowanceSufficient $minAllowanceNeeded]} {
   set next [lindex $urlsToVisit 0]
   set urlsToVisit [lreplace $urlsToVisit 0 0]
   # next: the next site/path/file to be visited

      regexp {([^/]*)(.*)} $next dummy currentSite currentFile
     # the file (incl. path) and the site are extracted
     lappend filesRead [ParseFileName "$currentSite$currentFile"]
     # mark the file as read

    if {![catch {ara_go $currentSite}]} {
     # successfully changed to the current site

     puts "Search agent [ara_me] arrived at $currentSite"

     lappend filesHere $currentFile

     set servicePoint [ara_meet "Document Retrieval Service"]
```

```
while {    [llength $filesHere]
       && [AllowanceSufficient $minAllowanceNeeded]} {
  # there is still a file at this site not examined, and
       there are
  # enough resources left to examine at least one file

  set currentFile [lindex $filesHere 0]
  set filesHere [lreplace $filesHere 0 0]
```

At this point, a specific document, identified by its path name stored in `currentFile`, has been chosen for examination. The search agent retrieves the content of this document by requesting it from the service point of the retrieval service. If the document could be retrieved, it will be examined and searched further:

```
set document [$servicePoint $currentFile]
lappend filesRead "$currentSite$currentFile"
if { $document != "Illegal operation: File does not exist" &&
     $document != "Illegal operation: Permission denied" &&
     $document != "Out of memory"} {
     SearchDocument $searchExpr
} else {
     puts "\"$currentFile\": $document"
}
unset document
```

The variable holding the document content is explicitly deleted (`unset`) after processing the document. This is a typical optimization measure prior to a migration, in order to avoid taking along data which is not really needed any more. If this deletion were omitted, the variable would be implicitly deleted by later overwriting it with the content of the next retrieved document; this, however, might not happen until after the next migration.

In order to reduce multiple visits to the `currentSite`, the agent examines all candidate documents there before moving on to another site, no matter what sequence they had been added to `urlsToVisit`. Note, however, that multiple visits cannot be avoided in the case where the agent learns of the existence of a candidate document only after having visited that document's site. The references to local files are thus transferred from `urlsToVisit` to `filesHere`:

```
while {[set further \
        lsearch $urlsToVisit "$currentSite/*"]] >= 0} {
  regexp "$currentSite\(.*)" [lindex $urlsToVisit $further] \
        dummy file
  set urlsToVisit [lreplace $urlsToVisit $further $further]
  lappend filesHere $file
}
```

```
      } ;# filesHere
```

When there are no more documents known to be searched at the local site, the agent leaves for the next site, and continues its loop until all references have been processed:

```
        $servicePoint leave
        puts "Search agent [ara_me] leaves $currentSite"
      } else {
        puts "Search agent cannot go to $currentSite"
      } ;# if go
    } ;# while urlsToVisit
  }
```

Processing an individual document is performed by the `SearchDocument` procedure. This extracts all hyperlink references from the document, and each reference, local or remote, is transformed into a normal form: The `http://` protocol specification is removed,[36] as well as any interspersed `//`, `/../` or `/./`. Local references are transformed into absolute path names, while global ones have an additional site name added at the front. Note that local references are not added to the general `urlsToVisit` list, but collected in the local `filesHere` list, as expected by the `SearchWeb` procedure in its effort to search local files before remote ones:

```
  proc SearchDocument{searchExpr} {
    global document
    global filesHere
    global filesRead
    global currentSite
    global currentFile
    global urlsToVisit

    if {![regexp $searchExpr $document]} {
      return
    }
    if {![GetTitle]} {
      puts "\"$currentFile\": Illegal format: No HTML-<TITLE>"
      return
    }

    if {![regexp {(/+.*/)([^/]*)} $currentFile dummy path]} {
      set path /
    }
    # Initialises path as the path of the currentFile
```

36 For the sake of simplicity, the client searches only for HTML documents accessible by the HTTP protocol (i.e. having URLs beginning with `http://`).

```
while {[regexp -nocase \
    "<A( |\t|\n)+HREF( |\t|\n)*=|\t|\n)*\"(\[^<\]*)\"( |\t|\n)*>" \
     $document dummy dummy dummy dummy reference]} {
  # while a new reference is found in the currentFile

  regsub -nocase \
    "<A( |\t|\n)+HREF( |\t|\n)*=( |\t|\n)*\"(\[^<\]*)\"( |\t|\n)*>" \
         $document "" document
  # deleting the reference in document

  set url [regexp -nocase "http://$currentSite" $reference help]

  if {$url || ![regexp -nocase {http://} $reference help]} {

  # it is a reference to this site (search for http-references
          only)

    if {$url} {
      # extracting the reference (including the complete path)
      regexp -nocase "http://$currentSite\(.*\)" $reference help \
            reference
    }

    if {![regexp {^/} $reference dummy] }  {
      set reference $path/$reference
    }

    # the path is added to the new reference if it is a
          relative one

    set reference [ParseFileName $reference]

    if {[lsearch $filesRead \
              [ParseFileName "$currentSite/$reference"]] < 0 &&
        [lsearch $filesHere $reference] <0} {
      lappend filesHere $reference
    }
    # if the new file has not been read before it is appended
          to the
    # list filesHere

  } else {
    # it is a reference to a remote site

    regexp {[^:]*://(.*)} $reference dummy reference
    set reference [ParseFileName $reference]
    # format of reference is now site/path/file
    if {[lsearch $filesRead $reference] < 0 &&
      [lsearch $urlsToVisit $reference] < 0} {
      lappend urlsToVisit $reference
    }
  }
```

```
  } ;# while
}
```

The search agent is completed by four auxiliary procedures for printing out the table of collected document references, parsing a file name, collecting the title of an HTML document, and checking whether the residual allowance of the agent is still sufficient. Note that file name parsing is performed here not for reasons of security, as in the server agent, but to arrive at a unique representation for each URL in order to prevent multiple searches of one document referenced by several equivalent, but not identical URLs. The source code of the procedures follows here:

```
proc PrintResult {} {
  global table
  foreach title [array names table] {
    puts "Title: $title"
    puts "Files: http://$table($title)"
    puts  ""
  }
}

proc ParseFileName {file} {
  while {[expr [regsub -all {/\./} $file / file] \
             + [regsub -all {//} $file / file] \
             + [regsub -all {([^/]+)/\.\./} $file / file]] } {}
  return $file
}

proc GetTitle {} {
  global document
  global table
  global currentFile
  global currentSite
  if {[regexp -nocase {<TITLE>([^<]*)</TITLE>} $document
          dummy title]} {
```

Enter the reference leading to this document into the table of document references. Note that this table associates document titles with a *list* of references, since there might be several copies of the same document at different locations, and the table is to list all of them. In any case, two documents bearing the same title are assumed to be identical:

```
    lappend table($title) ParseFileName $currentSite/$currentFile]
    return 1
  }
  return 0;
}

proc AllowanceSufficient {minAllowanceNeeded} {
```

```
    global sufficientTime
    global sufficientMemory

    set sufficientTime [expr [expr [lindex \
            [lindex [ara_get_allowance] 1] 0] == -1] || \
        [expr [lindex [lindex [ara_get_allowance] 1] 0] > \
            [lindex $minAllowanceNeeded 0]]]

    set sufficientMemory [expr [expr [lindex \
            [lindex [ara_get_allowance] 1] 1] == -1] || \
        [expr [lindex [lindex [ara_get_allowance] 1] 1] > \
            [lindex $minAllowanceNeeded 1]]]
    return [expr $sufficientMemory && $sufficientTime]
}
```

5.6 Ara system architecture and implementation

This section explains selected internal concepts of the Ara system which are useful to gain deeper insights into its rationale and capabilities, and also to combine it with new software components. The material presented here is for the technically interested reader, but not required for using the system or understanding the previous sections.

5.6.1 Processes and internal architecture

section 5.2 already introduced some basic concepts of the Ara architecture, namely agents, interpreters, and the core. Agents run within interpreters for their respective programming language, controlled and served by the common system core. It is a fundamental principle of Ara to execute agents as autonomous, concurrent processes. This supports their independent and possibly asynchronous execution, provides flexible control, and facilitates mutual protection.

As the system needs to control the agents in a rather fine-grained manner, it does not rely on the comparatively heavy-weight process abstractions and relatively coarse control usually provided by the host operating system. Furthermore, common process implementations, be they light or heavy, tend to be particularly platform-specific.

Instead, the Ara core provides its own process abstraction as the basis for agent implementation. Within such an *Ara process*, there is usually some interpreter running, processing the program of a mobile agent. However, from the point of view of the core, all processes are treated uniformly.

The core governs the agent processes, and mediates their interaction and host system access. Basic functions, such as migration, are provided to the agents by the core, while higher-level services are offered by server agents. While mobile agents are interpreted for reasons of portability and security, stationary agents may also be compiled, as discussed in section 5.4.9. Examples occur among the system processes: In addition to the agents visible at the application level, the Ara system also employs processes for certain internal purposes, in order to modularize the architecture. If such system processes are stationary, they are compiled. The process structure thus supports architecture modularity, without harming performance in any significant way.[37] An example of this is the communication process which handles the network interface of the host system. System processes have special permissions and may directly access the host operating system, bypassing the core. As the process implementation is internal to the Ara system, the complete ensemble of agents, interpreters, and core runs as a single application process on top of an unmodified host operating system, which considerably facilitates porting to specific platforms. Figure 5.10 depicts this basic architecture of the Ara system as a refinement of Figure 5.1.

Ara processes are implemented as individual threads of control within a common physical address space,[38] as provided by the host operating system. The core contains an efficient threads package, executing the threads concurrently in a nonpreemptive way. Synchronization primitives such as messages and semaphores are employed for internal purposes; application agents, however, are presented higher-level facilities such as service points. The common address space allows highly efficient process management and

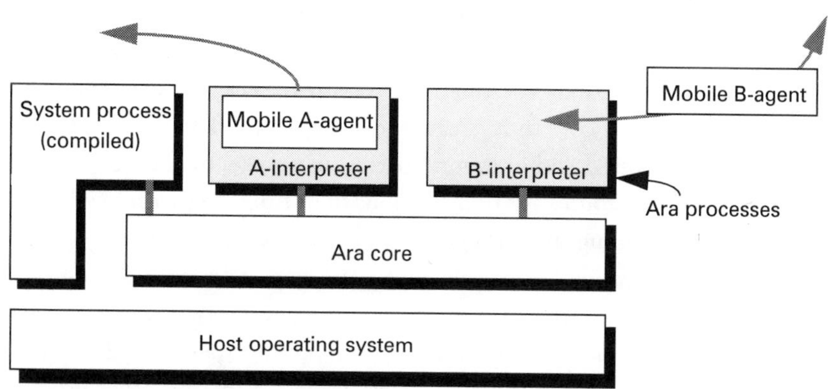

Figure 5.10 Ara system architecture

37 See the subsequent discussion on thread implementation.

38 Note that the interpreted agents are completely protected from each other nevertheless (see section 5.6.3, "Protection").

interaction. Thanks to the nonpreemptive process scheduling, there is no need for synchronization in the core and in the interpreters, further benefiting performance.

The reader may have noticed an apparent contradiction between the above description of process scheduling as nonpreemptive, and the statement of "Agent scheduling" on page 124, that the core enforces time sharing between the agents by preemption. This seeming contradiction is resolved when discriminating between processes, which execute machine code at the processor level, and interpreted agents, which interpret a program at the language level.[39] From point of view of the agent, execution is preemptive, since control may be withdrawn from the agent between any two primitive instructions in its program. This is implemented by a core function for time slice surveillance, which is called by the interpreter at some regular interval (after every Tcl primitive, in the case of a Tcl agent). This function performs a voluntary, nonpreemptive process switch whenever it finds the agent's time slice exhausted. This concept provides the preemption required for the execution of untrusted mobile agents, without introducing the complications of asynchronous processor contexts.

As the time slice surveillance function is called at a regular interval, this interval is also utilized to serve as the time quantum for the logical clock governing the timing service explained in the section on "Timing" on page 125.

5.6.2 The communication process

The communication process (comm-process) is the most prominent system process, handling the network interface of the host system for the purpose of agent migration. It accepts outbound agents in the form of a linearized byte array (agent parcels) and sends them to the comm-process at the parcel's destination site. Conversely, it receives inbound parcels and passes them on to the core to re-create the living agent. The comm-process can handle any number of inbound and outbound parcels in parallel by interleaving the individual transmissions. This accounts for transmissions to or from sites with low-bandwidth connectivity without harming the throughput on high-speed connections.

Remembering the discussion in "Agent scheduling" on page 124, on blocking I/O operations, the comm-process currently emulates a proper process-aware network access. To this end, it uses nonblocking network operations in a loop interspersed with releases of control by retiring whenever the access would block. Provided that no other process retains control overly long, this realizes the desired behavior of network I/O in parallel with other agents executing, at the affordable price of some unnecessary network polling.

---

39 In the case of compiled agents, the level of interpretation does not exist, and scheduling is indeed nonpreemptive even from point of view of the agent.

Agent parcels are transmitted as a single, unidirectional data packet from the source to the destination site comm-process. The transmission may be acknowledged, if the concrete transport protocol supports this, but this is not required. At the time of this writing, the comm-process uses TCP as its only transport protocol, with an optional gateway to the AX.25 radio transmission protocol.[40] However, further protocols such as SMTP (email) will be added, and it is planned to let the comm-process choose a protocol for each parcel transmission adapted to the current connectivity, provided there is more than one alternative available. In any case, the actual choice of protocol will not be visible to the migrating agent.

5.6.3 Protection

Mobile agents executing within an Ara system are protected from each other, as is the core, in order to prevent malfunctioning or malicious agents from spying or tampering outside their own boundary. Protection in this general context means control of the data read and written and the external functions called by an agent; higher-level issues of object access authorization, such as entering a place, or fetching requests from a service point, are treated specifically.

Since mobile agents are interpreted, and since an interpreter obviously has complete control over the interpreted program, protection can be achieved independent of hardware facilities like privileged processor modes or page protection. Concerning the functions called, protection is trivial, since every call has to pass through a stub defined by the interpreter which can be trusted because it is out of the agent's reach. Data access protection is achieved through an address space concept: borrowing from operating systems terminology, the set of data visible to an agent might be viewed as the agent's address space—both if the data exists in the form of a randomly addressable memory area (as in C), or in the form of a set of unordered symbolic variables (as in Tcl or Java). The interpreter ensures that each agent has direct access to its own address space only, while core functions must be used to interact with other agents and the core. The core and its objects, such as (other) agents and service points, exist outside the agents' spaces in real memory.

Confining access to an agent's own address space is trivial in the case of exclusively symbolic variables (since each interpreter maintains its own set of these). If, on the other hand, the interpreted language allows random memory accesses, the interpreter must provide a virtual memory image to the program using some kind of address checking and translation. The MACE interpreter for the C language is an example of this.

---

40 This is used for experimentation within the Ara project, and is not included on the CD-ROM.

In any case, agents need a means to name core objects existing outside the agent's address space. The object IDs already mentioned in section 5.4 provide this means, in addition to their purpose of uniquely naming the core objects. Object IDs have a representation in the agent's language (e.g. a string in Tcl) which can be viewed as an opaque pointer into core space. When a core function is called by an agent, argument objects are supplied in the form of IDs which are mapped to their objects by the stub for this function, using an object table maintained by the core. Again, this is reminiscent of operating systems such as UNIX, where applications name kernel objects such as files indirectly through IDs, although not usually globally unique ones.

The stubs also perform any interpreter-specific parameter checking, such as checking addresses for containment in the program's address space. This mapping and checking work has been placed in the stubs rather than in the core functions themselves in order to avoid unnecessary work when a core function is called from a trusted context, such as from the core itself.

5.6.4 Saving and restoring the state of a process

At the heart of mobile agent implementation in Ara is a mechanism to save the state of a process and restore it after transportation to another site. As explained in section 5.2.3, "Agent mobility: going from place to place," Ara agents can move without interfering with their execution. They continue from the same state in their program which they had reached when leaving the source system. This orthogonal migration implies that their execution state has to be extracted from the source system and reinstalled into the destination system as illustrated in Figure 5.11.

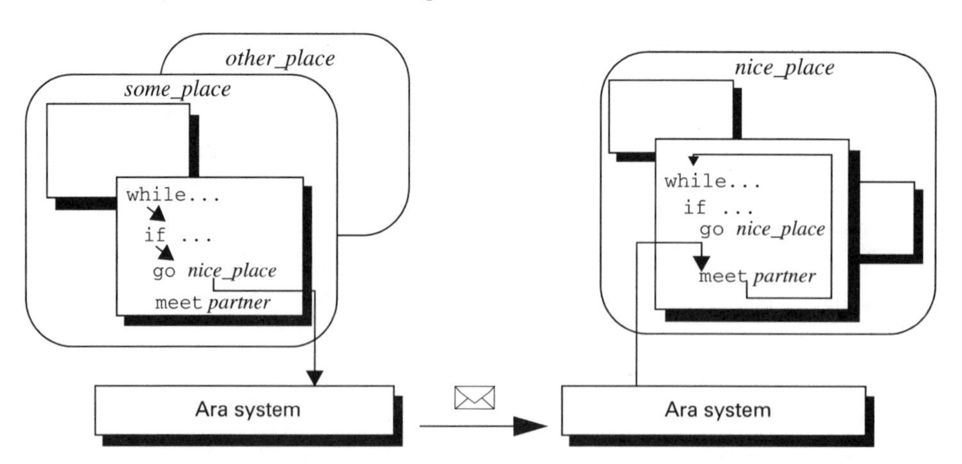

Figure 5.11 Orthogonal migration between Ara systems

Such an extraction and reinstallation is a delicate procedure, the more so when the participating machines have different architectures, since the execution state of a process generally exists in a machine-dependent format such as assuming a specific word length or byte order. In Ara, the state of an agent is transformed into a portable form prior to moving. This transformation is facilitated by the Ara system architecture: an agent's state is composed of the state of its specific interpreter on the one hand, and the state of its underlying general Ara process in the core on the other hand.

The core performs the portable transformation of the Ara part on its own, whereas it uses a dedicated function defined by the interpreter (upcall) for the other part. This function may build its implementation on a number of utility functions offered by the core. This provides the transformation of common data types into a portable form, as well as a general concept to transform the states on the individual levels of a procedure-call hierarchy on the run-time stack. The run-time overhead of the complete transformation into the portable form (remembering section 5.6.2, this is called parceling a process) depends on the complexity of the program state at the time of migration, and on the general complexity of the interpreter's state representation.

Interpreters for most interpreted languages are implemented in a procedural language (in C, most of the time). In procedural languages, the execution state of a running program is to a large part contained in the program's run-time stack at a given instant. This causes a problem with the normal implementation of interpreters when trying to save and restore the interpreted program's execution state in a portable way. The normal interpreter implementation intertwines that state with the interpreter's run-time stack, the latter being invariably machine-dependent. There are basically two solutions to this problem: Either reimplement large parts of the interpreter, completely replacing the use of the run-time stack by a new scheme,[41] or continue as before during normal execution, but transform the run-time stack into a portable representation when state saving or restoring is needed.

The primary design considerations of Ara's scheme for saving and restoring execution states were efficient overall execution and sufficient generality as to be applicable to any software implemented in C, while keeping that application so straightforward that it can be easily extended to unknown software (e.g. interpreters for additional languages) without understanding its internal workings. For these reasons, Ara adopted the latter of the above alternatives; a portable transformation of the current run-time stack at migration time. The devised scheme consists of an annotation of the source code of a given interpreter implemented in C, which requires only a tightly

41 This usually involves some explicit substitute for the run-time stack, as is done, for example, in a stack machine.

localized[42] understanding of the source, can be automated to a large part, and adds no measurable penalty to normal execution speed. This allows additional language interpreters to be adapted to the Ara core with reasonable effort. The adaption procedure has been applied so far to the Tcl interpreter[43] and the MACE interpreter, and is currently being applied to the Java interpreter.

5.6.5 Cloning and checkpointing

Both cloning and checkpointing involve saving the state of a live process and restoring the process from this. Cloning restores a copy of the saved process immediately, while checkpointing defers restoration until explicitly demanded. The problem of saving and restoring a process's state, however, has been solved to implement migration in the first place. It should come as no surprise, therefore, that the implementations of cloning and checkpointing are mostly based on the migration implementation. In fact, their realization was remarkably simple by falling back on migration. In a sense, a process cloning itself is realized as a process migrating to the local place, but being duplicated on arrival. Analogously, checkpointing is implemented like a process migrating to the disk.

There is one drawback with this implementation. Processes migrating away from the local system leave behind their external state (see section 5.2.3, "Agent mobility: going from place to place"); this also applies to processes created by cloning and those restored from a checkpoint. For the case of cloning this is unfortunate, since the objects of the process's external state still exist unchanged after cloning, so that cutting the clone's relations to them is not really necessary. It is possible that the cloning implementation will be extended accordingly in a future release. The case with checkpointing is different since the objects of the process's external state might indeed change arbitrarily between checkpointing and restoration, so there is no point in trying to preserve the external state.

5.6.6 Adaption of further programming languages to Ara

One of the central motivations for the separation of language interpreter and system core in the Ara architecture was the desire to add further agent programming languages as they seem useful. To this end, the language interpreter must be adapted to Ara in

42 In nearly all cases, limited to one C function.

43 For the purpose of validation, an automatic saving and immediate restoring of the program after *every single* program step can be arranged. The modified Tcl interpreter passes this validation for the complete official Tcl test suite without an effect on its function.

various respects. This adaption, sketched in Figure 5.12, is a well-defined and straightforward part of Ara.

In contrast to the interpreted agent, the interpreter itself is conceptually a part of the Ara system, is trusted, and supports the core in its tasks pertaining to the interpreted agent. The duties of an interpreter in this respect include the definition of calling interfaces (i.e., stubs) in its programming language for the functions imported from the core, and conversely to provide functions for interpreter management (upcalls) to the core. The work of the stubs is mostly a matter of data format conversions and similar interface translations; in particular, they must map between object IDs for core objects and the objects themselves, as described in section 5.6.3, "Protection."

Regarding the interpreter upcalls, the functions for the portable state extraction and restoration described in section 5.6.4 are the most prominent here; others include system and interpreter creation and deletion, and a function to start the interpretation. A general requirement for the interpreter is to satisfy demands for dynamic memory arising during the interpretation exclusively by the memory management functions provided by the core (the same as called by the stubs of section 3.11).

This ensures that any memory consumption on behalf of the interpreted agent is properly accounted to its allowance. Further, the interpreter must possibly provide a virtual memory image as described in section 5.3. Finally, the interpreter has to assist the core in preemption, performing regular calls to the core function for time slice surveillance, as described in section 5.6.1, "Processes and internal architecture."

In principle, any interpreted programming language can be adapted to Ara: a given interpreter must be extended by mechanisms to serve the described functions. The main expense in the adaption of a concrete interpreter is usually spent in the implementation of the state extraction and restoration, the complexity of which directly depends on the

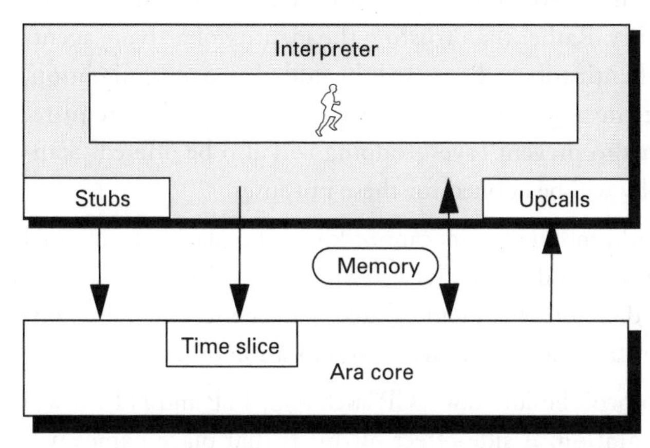

Figure 5.12 Adapting a language interpreter to the Ara core

complexity of the state representation in this interpreter. While space does not permit a detailed real language adaption here, it is instructive to look at an exemplary adaption of an ad-hoc interpreter for some toy language. The software distribution on the CD-ROM contains the source code for such a toy interpreter adapted to Ara. Note that programmers adapting a new language interpreter to the Ara core will need to use additional core functions, exceeding the application functions covered in section 5.4, "Ara programming concepts and features." For information on these, please refer to the software distribution on the CD-ROM, where they are collected in a declaration file.

5.7 What may be expected from Ara in the future?

The Ara system is in the midst of development, and besides completion of those features described, but noted as unfinished in this text, further concepts will extend the system. The following outlook names some fields where work is in progress or scheduled to begin.

- As the very first step, the MACE system for the implementation of fully mobile agents in C will be included in the Ara software distribution. It is possible that this will have happened by the time of publication of this book. Please check the WWW page for Ara on the CD-ROM.
- The most prominent new concept will be a secure and portable host interface; this was discussed in section 5.2.4.
- The power of a mobile agent as conceded by a specific host system crucially depends on the agent's identity. Rather than trusting the identity of arriving agents on good faith, a secure authentication scheme will be added to agent migration, optionally authenticating the agent or the receiving system as they require. Encryption of moving agents to prevent eavesdropping will also be offered. Standard public-key cryptography will be applied for these purposes.
- In a realistically sized network, mobile agents cannot know the places of potential relevance to their task in advance. A directory service will be provided for this, mapping services to places. The directory will be integrated with the service point concept by optionally publishing a service point's name and location on creation.
- Further communication protocols besides raw TCP, such as SMTP and HTTP, will be added as options for migration. A side-effect of this is that place names will become more structured, which is additionally furthered by allowing more than

one place per Ara system. Place names will resemble a collection of URLs, one for each communication protocol by means of which the place can be reached.

- Java, the new interpreted programming language, offers interesting concepts for mobile code and enjoys rapidly increasing importance in networked programming. The Java interpreter will therefore be adapted to the Ara core, as a third language besides C and Tcl, to enable mobile agents programmed in Java.
- The performance of the MACE bytecode interpreter Ara uses for mobile agents written in C, although remarkably good for an interpreter, will be further improved by compiling the bytecode to the native machine code of the host system at run time. Security checking code inserted by this compiler will ensure that this native code will satisfy the same security requirements and interact with the Ara core in the same way as an interpreter would do.
- World Wide Web support will be added to Ara, both at the user and the system levels. For instance, the system will be equipped with a user interface through a Web browser, and server agents will provide Web access to application agents.

PART III

A Java-based agent system

This section presents a mobile agent system developed by IBM Tokyo, Aglets Workbench, and its associated software. Aglets are based entirely on the java language and have features and limitations based on the use of Java

Java presents a complete system built for mobile code, based on the concept of a virtual machine; software developed in Java should be able to run on any machine that is running a Java virtual machine. This promises a dramatically different agent server model than that of the systems presented in Part II.

At the same time, Java was designed to be the optimal language for programmers desiring speed, portability and security. For solutions needing a solid implementation of each of these features, Java presents even more of an advantage over other languages and systems. But in providing a complete solution, Java actually is limited to the extent that each of the features can be extended to compete with other languages. As an example, Java is slow compared to C or C++ and will remain so in its present vision; Java is often interpreted while C and C++ are compiled. But C and C++ have no security model, whereas Java provides protection against flawed code and the introduction of viruses into a system.

chapter 6

Aglets workbench

CO-AUTHORED BY THE IBM AGLETS WORKBENCH TEAM

Since its inception, Sun's Java language has been promoted as the panacea for all computer networking problems. Over the past year, as the language has grown and developers have begun to deploy solutions based on Java and its related technology, some of the initial promises have begun to appear.

IBM Tokyo Research Laboratory is one of the first groups to provide a comprehensive Internet agent system based on Java. This chapter takes an initial look at the work IBM has done and presents a number of simple examples based on their work.

Java provides a number of new features to agent developers, as well as providing a number of new constraints. Three agent systems in this book utilize Java as their agent language; Ara (not yet completed), Aglets, and JavaAgentTemplate. Each utilizes Java in a slightly different manner and can be used to compare different approaches to the agent problem under the same constraints. Ara implements Java agents in terms of its own architecture which was designed to be language independent. The Ara approach to Java is much different than that of the other two systems due to its focus on agents first, language second.

6.1 Aglets

Aglets is one of the first complete Internet agent systems to be developed on the Java class library. Aglets were developed by the IBM Tokyo Research Laboratory.

Aglets Workbench is a visual environment for building network-based applications that use mobile agents to search for, access, and manage corporate data and other information. Aglets Workbench allows users to create mobile platform-independent agents based on the Java programming language. Its visual builder allows you to quickly compose personalized agents that can roam the Internet, and the rich set of software components in the Aglets Workbench enables agents to access corporate data bases, search, travel, and communicate in a standardized and secure manner.

6.2 Aglets Workbench

Aglets Workbench provides the ability to create Java-based programs that can be dispatched from one computer and transported to a remote computer for execution. Arriving at the remote computer, the Aglet agents present their credentials and obtain access to local services and data. The remote computer may also serve as a broker by bringing together agents with similar interests and compatible goals, thus providing a meeting place at which agents can interact.

Aglets Workbench is so powerful because it allows users to create mobile platform-independent agents based on the Java programming language. This is the first industry-supported agent system that has platform-independence as a design goal.

Its visual builder allows you to compose personalized agents that can roam the Internet. The set of software components in the Aglets Workbench enables agents to access corporate data bases, search, travel, and communicate in a standardized and secure manner.

6.2.1 A bit of Java

There is an overabundance of Java books on the market so we are not going to attempt to present the language itself. Instead we will focus on the one feature of Java that makes it the ideal language to implement mobile agents on the Internet.

The one feature of Java that makes it distinct from most other languages, and which provides one of the major features of future agents, is that applications written using the basic libraries are guaranteed to be portable across platforms. Once the code is written, it never needs to be recompiled or ported in order to get it to run on different platforms. The reason this is so important is the large number of different operating systems that make up the Internet. Macintosh, Windows 3.1, Windows 95, and UNIX are only a few of the operating systems that an agent could possibly visit. But an application written in C or C++ for one of these operating systems (and likewise, one agent) cannot run on all.

Java has solved this problem.

This is because the Java compiler on any one of the platforms actually produces a platform-neutral application. The Java runtime environment on each of the machines (called a Java virtual machine) translates the downloaded neutral code into machine-specific instructions. Java virtual machines are being built into most browsers or into the operating system. Agents written in Java will be turned into this neutral code before they are released on the Internet. At each machine, the agent would be translated into the local machine code and could operate on that machine. When the agent is ready to move on it can package the information it has obtained and move to another platform, eventually making its way back to its home machine.

6.2.2 The aglets framework

The cornerstone of Aglets Workbench is the aglets framework. This framework, which is Java based, provides the mobile-agent-specific components of the workbench. It also introduces the notion of an *aglet*, a mobile (agile) agent written in Java and named after its framework base class.

Using the Aglet class is a convenient way for user-defined agents to inherit default properties and functions for mobile agents. These include:

- A globally unique naming scheme for agents.
- A travel itinerary for specifying complex travel patterns with multiple destinations and automatic failure handling.
- A white board mechanism allowing multiple agents to collaborate and share information asynchronously.
- An agent message-passing scheme that supports loosely coupled asynchronous as well as synchronous peer-to-peer communication between agents.
- A network agent class loader that allows an agent's Java byte code and state information to travel across the network, and an execution context that provides agents with a uniform environment independent of the actual computer system on which they are executing.

Another way in which the framework supports the agent developer is through the introduction of usage patterns. Pattern-based design and development has proven very successful in a number of areas. We have added a number of high-level agent usage patterns to the framework. These usage patterns describe common dual relationships between agents such as master-slave, messenger-receiver, and notifier-notification. These patterns have been represented by a number of classes that can be used as templates by the agent developer.

The comprehensive aglets framework is entirely written in Java to ensure maximum portability. Its class APIs are fully documented, allowing mobile agents to be directly integrated into existing and future network-based corporate information systems.

6.2.3 A visual development environment

Visual builders are essential for programmer productivity. Using a drag-and-drop metaphor, the visual builder in Aglets Workbench allows you to visually combine graphical as well as nongraphical components into personalized mobile agents. The visual builder named Tazza makes it easy for the agent developer to create agent applications complete with mobile behavior and visual front ends.

6.2.4 *Accessing corporate data*

Access to corporate data bases will be essential to many mobile agent applications. Aglets Workbench offers several packages for data access, including JDBC/DB2 and JoDax. The packages and their usage are also included on the CD-ROM.

6.2.5 *The agent transfer protocol (ATP)*

The Agent Transfer Protocol (ATP) is used to transfer agents over the network. ATP is an application-level standard protocol for distributed agent-based information systems. Aimed at the Internet and using the Universal Resource Locator (URL) for agent resource location, ATP offers a uniform and platform-independent protocol for transferring agents between networked computers.

While mobile agents may be written in many different languages and for a variety of vendor-specific agent systems, ATP offers the opportunity to handle agent mobility in a general and uniform way. For example, any agent host machine will have a single and unique name independent of the set of vendor-specific agent systems it supports. ATP also provides a uniform agent transport mechanism and allows a standard agent query facility to be used throughout the network.

The first version of ATP, named ATP/0.1, is implemented as an independent and fully documented package in the Aglets Framework.

Entirely written in Java, this highly portable set of classes provides a standard API for creating ATP daemons, connecting to ATP sites, and generating ATP requests and responses. This package is independent of any particular agent implementation.

6.2.6 *Visual agent manager*

Tahiti is a visual agent manager based on the aglets framework. Tahiti uses a unique graphical user interface to monitor and control aglets executing on your computer. Through a drag-and-drop interface you can make two aglets communicate, or dispatch an aglet to a particular site. Tahiti is more than a system administration tool; it is a desktop tool for agent users in the same way that the Web browser is become the fundamental tool for Web users.

6.2.7 *Enabling aglets on the Web*

The World Wide Web is becoming a very important infrastructure for mobile agents. Agents developed on Aglets Workbench can be embedded in Web pages in a way that allows clients to launch these agents on the Internet directly from their Java-enabled Web browsers.

Aglets Workbench includes an agent Web launcher named Fiji. Fiji is a Java applet based on the aglets framework and therefore capable of creating an aglet or retracting an existing aglet into the client's Web browser. The Fiji applet simply takes an agent URL as its parameter and can easily be embedded in a Web page by using HTML, like any Java applet. And just as with other applets, all the required software will be dynamically downloaded to the browser as it is needed. The client will not have to explicitly download and install Aglets Workbench in order to use your agents.

This is a way to empower your Web pages with mobile agents.

Web sites can also be empowered with mobile agents. If the Web server is paired with an ATP daemon supporting aglets, Fiji applets will be able to dispatch aglets to the Web site for remote search or disconnected Web site monitoring. We envision a whole collection of Web service agents for aglets-enabled Web sites.

6.2.8 Just how secure are aglets?

Security is of paramount importance to users of mobile agents. On the one hand, mobile agents provide great power that can be used to achieve very valuable ends; on the other hand, they could be subverted to become a breeding ground for software viruses. Receiving unknown agents from across the network is potentially an open invitation to all sorts of problems.

The aglets framework supports an extensible layered security model. The first layer comes from the Java language system itself. Imported code fragments in agents are subjected to a series of checks, starting with tests to ensure that the code format is correct, and ending with a series of consistency checks by the Java bytecode verifier.

In the next layer is the security manager, which allows users of the aglets framework to implement their own protection mechanisms.

Our foremost challenge and top priority in making Tahiti has been to construct a system that will guarantee secure hosting of agents. Tahiti implements a configurable security manager that provides a fairly high degree of security for the hosting computer system and its owner. The default security configuration is very restrictive. Any attempt by an agent to access a file to which access has not been granted will be regarded as a security violation, and the agent will not be allowed that specific access.

In the third and final layer is the Java security API, which is a framework that makes it easy for agent developers to include security functionality in their agents. This functionality includes cryptography, with digital signatures, encryption, and authentication.

6.2.9 *Standardization efforts*

We are actively participating in agent standardization efforts. More specifically, the ATP and the framework API of Aglets Workbench has been submitted to OMG as a proposal for OMG's mobile agent facility.

Furthermore, the ATP/0.1 specification is a public document with unrestricted distribution. The protocol has been modeled over HTTP, and proposes a standard way of transporting agents independently of any particular implementation.

6.2.10 *Making aglets ubiquitous*

IBM has made it a top priority to make aglets ubiquitous.

IBM has made Aglets Workbench freely available for downloading at sites in the U.S. and Japan, and has allowed the package to be included with the CD-ROM. IBM has introduced a download now logo program, and has endorsed mirror sites at universities and other organizations. IBM is also working on efforts to bundle Aglets Workbench with Web and database servers to ensure that aglets are readily available. Finally, IBM is working on an aglets enabled logo program for Web sites hosting an ATP daemon that will accept aglets.

6.2.11 *Aglets as a solution*

The Aglets Workbench from IBM offers a first-of-its-kind visual environment for building network-based applications using mobile agents. At the time of writing, the workbench includes the following packages:

- The Aglets Framework for mobile Java agents
- The Agent Transfer Protocol (ATP)
- Tazza—a visual agent builder
- JDBC for DB2
- JoDax for corporate data access
- Tahiti—a visual agent manager
- Fiji—an agent Web launcher

Aglets Workbench provides a new and viable paradigm that unifies mobile and stationary objects; object-passing, message-passing, and data-passing; autonomous and passive objects; asynchronous and synchronous processing; local and remote objects; disconnected and connected operations; and parallel and sequential execution.

More information is on the CD-ROM along with all of the examples presented below and pointers to the continuing development going on at IBM.

6.3 Getting started

It is extremely easy to run aglets. Below are the steps needed to get your first aglet agent running on the Net.

Install the aglets software

The first thing you need to do is insert the CD-ROM and install the aglets library onto your PC. Installation is explained in detail in the IBM Aglets section of the CD-ROM.

Download the Java software

If you do not already have JDK 1.0.2 and RMI (object serialization) you should download and install those packages, too. The Aglet library is based on the Java Development Kit (JDK), which is being revised by Sun Microcomputer's Javasoft division at a rapid pace. The reader should refer to the Javasoft Web site `http://www.javasoft.com` to obtain the latest JDK along with the latest RMI (Object Serialization).

Having successfully downloaded and installed these three items you are ready to enter the world of mobile Internet agents.

Run two ATP daemons

Warning: The pre-release version on the CD-ROM is equipped with a limited security manager.

Create a DOS prompt window and start the daemon in the ATP package:

```
java atp.AtpDaemon -ap Aglets
```

This daemon will listen for agents on the default port 434—a port reserved for mobile agents.

Create another DOSþprompt window and start another daemon here:

```
java atp.AtpDaemon -port 500 -ap Aglets
```

This daemon will listen for agents on another port—in this case port 500. You can actually choose any port which is currently not occupied by some other application. It also makes Tahiti appear on your screen. Tahiti is a desktop interface to manage running agents. The first time you run Tahiti it will ask you to register. Please do so to continue.

To get usage information about the command line options for the AtpDaemon type:

```
java atp.AtpDaemon -help
```

Start your first aglet

Now your first aglet should be running—look for it in Tahiti's window. Tahiti has been configured to automatically start the aglet named `aglets.demo.start.GoSay-Hello`. Use the mouse to highlight this aglet in Tahiti and push the Dialog button. The aglet will now display its own dialog window. The `GoSayHello` aglet is able to carry a message from you to any `atp.AtpDaemon` on the Internet. Type your message in the text area.

Dispatching your aglet

Now it is time to dispatch your aglet to the `atp.AtpDaemon` closest to you, the one listening to the default agent port. You should fill out the destination field in your aglet's dialog window. The destination for an aglet is a URL starting with the atp protocol name:

```
atp://your.host.com
```

Push the Go button on the aglet and it will dispatch itself to the specified destination. You will see another instance of Tahiti pop up with the dispatched aglet entering a few moments later. Watch for it to show your message in Tahiti's window.

Having delivered the message, the aglet will return to its origin and wait for new instructions.

Experiment

Now it is time for you to experiment with your aglet. We have some suggestions:

- Run `atp.AtpDaemon` on remote machines and play with GoSayHello.
- Look at source code for GoSayHello here.
- Take a look at the documentation for the `Aglet` class.
- The `AgletContext` class documentation may also be useful.
- Try some of the other cool demo aglets here.

6.4 The finger application

Finger is an aglet-based implementation of the more well known finger application in UNIX. It consists of two aglets: one stationary aglet (`Finger`) and one mobile aglet (`FingerSlave`). The latter is dispatched by Finger to a remote location where it will retrieve local user information.

Having gathered the local information, `FingerSlave` returns to `Finger` and delivers the information.

Finger is a particularly good example of two aglet usage patterns, namely the `Master-Slave` and the `Messenger` patterns. Continue below for a more detailed explanation of the use of these two patterns in the application.

Run Finger!

Run Tahiti

If you are not running Tahiti on your computer, now is the time to start it.

Create finger

Push the "New" button on Tahiti. Then you will find a panel for aglet creation. Please fill aglet name field with `aglets.demo.finger.Finger`. (Aglet name is case sensitive.)

You do not have to specify Source URL for `Finger` so leave that blank. When you have successfully created an `Aglet` instance, its class name will be registered in the AgletHotlist. You can later select it from that list. After setting Aglet name, push the "New" button and a Finger window appears.

Specification of the remote-server URL

Specify Which URL field with the target server's URL. The URL for the agent-server must start with `atp://`, followed by the host-and-domain address. You may specify the port number. Examples of URL are as:

```
atp://aglets.trl.ibm
```

or

```
atp://aglets.trl.ibm:434
```

(The preceding address is not a real one, please do not try it.) If you have ever successfully sent agents to any agent-servers, you will find the host names in URL Hotlist. Find the URL you want, fill in "Which URL," and press Go to create and dispatch an agent.

Getting the result

The user will receive the following information in the middle of the window.

1 Host Name: target server's host.domain name and the port number
2 User Name: registered user name in server
3 Organization Name: registered organization name in server
4 Email Address: registered organization name in server
5 Registered: registration date
6 Local Time: time of visit in local time in the textArea

The message from moving agents appears at the bottom of the window.

Troubleshooting

If there is only one item in URL Hotlist, you cannot get the selected item. You will get no reaction by pushing "Go" if you keep the "Which URL" field empty.

6.4.1 *Implementation*

The `Finger` class is inheriting the Aglet class and implementing two interfaces: `Master` and `Receiver`.

```
public class Finger extends Aglet implements Master, Receiver {
        ...
    }
```

The `FingerSlave` is inheriting the `Slave` class and thus indirectly also the `Aglet` class.

```
public class FingerSlave extends Slave {
        ...
    }
```

6.4.2 *Methods related to the master-slave usage pattern*

As a master, `Finger` should implement at least two methods that work as entry points for returning slaves. The first is `callback()` that should receive and handle the result of a slave's execution at some remote aglet server. In this example the slave returns with an instance of `FingerInfo` (a class for transferring data) and that string is passed to a display function by the `callback()` method.

```
public synchronized void callback(Slave s, Object arg) {
```

```
// Okay, my dear Slave is back...
setTheMessage("Finished");

// Displays the result of the search in a separate window.
_msw.setResult(((FingerInfo)arg).toTextBlock());}
```

The second method necessary to receive slaves is `inError()`. This method is called by the returning slave if it, for some reason, failed to perform its task. In this example `inError()` will display the error message from the slave:

```
public synchronized void inError(Slave s, Object message) {
      // Oh oh, that Slave failed to do its job...
      setTheMessage((String)message);}
```

Before the master can receive a slave, it must first create the slave. In this example the slave is created in a method called go. `go()` is invoked as a callback function from the GUI. The target server for `Finger` is specified by the user in a dialog window. The target is specified as a URL. The target is used along with other information in the construction of the slave which automatically will dispatch to its target. Consult the `Slave` class for more detailed information.

```
public void go(URLurl){
String _hostname=url.getHost();

setTheMessage("Starting...");

if(_hostname.equals(""))
     _hostname="localhost";

setTheMessage("tryingtogoto:"+_hostname);

//NowcreatetheSlave.Itisautomaticallydispatchedasapartof
//theMaster-Slaveusagepattern.
try{
     Slave.create(null, SlaveClassName, getAgletContext(),
                  this, newItinerary(url), newString());
  } catch(AgletExceptionae) {
     inError(null, ae.getMessage());}}
```

And now for the slave. The slave only has to implement one method called `doJob`. `doJob()` is an abstract method in the `Slave` class, and a slave is supposed to implement this method. In this example the slave's task is to retrieve local user information and keep it until it returns to the master. The protected variable `RESULT` is used to save the information in before returning to the master.

```
public synchronized void doJob() {
      RESULT = getLocalInfo();}
```

```
protected Object getLocalInfo() {
      AgletContext ac = null;

     try {
         ac=getAgletContext();
      } catch (AgletException ae) {
         return ae.getMessage();}

FingerInfo info =
   new FingerInfo (ac.getProperty("host-name", "Unknown host") +
                   " port: " + ac.getProperty("port", "7070"),
                   ac.getProperty("user-name", "Unknown"),
                   ac.getProperty("org-name", "Unknown"),
                   ac.getProperty("email-address", "Unknown"),
                   ac.getProperty("registered", "Unknown"),
                   (new Date()));
      return info;}
```

6.4.3 *Methods related to the messenger usage pattern*

The `Finger` class is also implementing the `Receiver` interface. This interface only requires a receiver to implement one method named `message`. `message()` is entry point for incoming messengers. In this example, messengers are dispatched from `FingerSlave` to `Finger`. `FingerSlave` creates a messenger as soon as it starts to perform its task at a remote server. Consult the `Messenger` class for more detailed information.

The messenger moves to the location of the receiver and hands over the message by calling `message()` in the receiver. The message is in this case passed on to the user.

```
public synchronized void message(Messenger m, Object message) {
      // Receives messengers sent by the Slave.
      setTheMessage((String)message);}
```

If you wish to study the `Finger` application in further detail you should consult the source code directly.

6.4.4 *Class finger: master aglet*

The first part of the source for this piece of the agent defines the Java classes that are going to be used and their interfaces. The class definitions are not included in the text but can be referenced on the CD-ROM.

Class `Finger` is used to retrieve local user information from a remote aglet server. This information includes the name, organization, email address, and the local time at the remote server. Using the master-slave usage pattern, the `Finger` class plays the role of the master. Given a URL (Agler Resource Locator) it will dispatch a slave (the `FingerSlave` class) to retrieve local user information. The slave will return with the information to be displayed by the master. While at the remote server, the slave will dispatch a progress report to the master. This is a part of the `Messenger` usage pattern. `Finger` is acting as a receiver in this pattern.

```
public class Finger extends Aglet implements Master, Receiver {

   // The main interaction window.
   private FingerWindow _msw = null;

   // Aglet web source (the URL & classname of the slave class).
   protected static final String SlaveClassName =
   "aglets.demo.finger.FingerSlave";
```

This makes this aglet immobile with `AgletException` if someone tries to dispatch this aglet.

```
public synchronized void onDispatching(URL URL) throws
     AgletException {
      // I will shout if you try to move me!
      throw new AgletException("Don't ever try to move me!");}

public synchronized void onDisposal() {
      // Removes any windows if disposed.
      if (_msw != null)
         _msw.dispose();}

public synchronized void onDialog() {
      if (_msw != null)
         _msw.show();}
```

This is the master's callback and the entry point for the returning slave. This method is a part of the master-slave usage pattern.

```
   public synchronized void callback(Slave s, Object arg) {
      // Okay, my dear Slave is back...
      setTheMessage("Finished");

     // Displays the result of the search in a separate window.
      _msw.setResult( ((FingerInfo)arg).toTextBlock());}
```

This provides the entry point for returning slaves that are in a state of error. This method is also a part of the master-slave usage pattern.

```
public synchronized void inError(Slave s, Object message) {
      // Oh oh, that Slave failed to do its job...
      setTheMessage((String)message);}
```

```
/** Sets the message line in the interaction window. */
   protected synchronized void setTheMessage(String message) {
      super.setMessage(message);
      if (_msw != null)
         _msw.setMessage(message);}

//--  clears the output areas in the main window.
private void resetTheWindow () {
   if (_msw != null) {
   _msw.clearMessage();
   _msw.clearResult();}}
```

This section implements the message interface of a receiver. This method is a part of the `Messenger` usage pattern.

```
   public synchronized void message(Messenger m, Object message) {
      // Receives messengers sent by the Slave.
      setTheMessage((String)message);}

/** The main entry point of the aglet's own thread */
public void run() {
      Vector v = new Vector();
      // Here we are going.
      setTheMessage("Starting...");
      try { _msw = new FingerWindow(this);
      } catch (AgletException ae) {
          inError(null,ae.getMessage());}}
```

Creates and sets up the slave with necessary information for it to dispatch to a remote aglet server and hopefully return safely. The parameter URL in `go(URL url)` represents the URL of the destination site.

```
public void go(URL url) {
       String _target = url.toString();

       setTheMessage("Starting...");

       if  (url.getHost().equals(""))
         _target="localhost";
       resetTheWindow();
       setTheMessage("trying to go to: " + _target);
```

We now create the `Slave`. It is automatically dispatched as a part of the master-slave usage pattern.

```
   try { Slave.create(null, SlaveClassName, getAgletContext(), this,
new Itinerary(url), new String());
   } catch (AgletException ae) {
           inError(null, ae.getMessage());}}}
```

This section presents the window classes for the interface that is presented for the aglet. The Class `FingerWindow` represents the main window for user interaction.

```
class FingerWindow extends Frame {

private static final String FONT = "Times Roman";
   private static final String SIZE = "12";
   private static final String TITEL = "Finger";
   private static final String MAL_FORMED_URL_MSG =
              "Invalid destination address.\n" +
                "Please type the correct destination to go.\n" +
                  "Example: atp://java.trl.ibm.com.";

private Finger _aglet = null;
private MessageDialog _helpWindow = null;

FingerWindow(Finger a) throws AgletException {
      super(TITEL);
      _aglet = a;
      int rgb = 0;
      int fs = Integer.parseInt(SIZE);
      setLayout(new BorderLayout());

try {
fs=Integer.parseInt(_aglet.getAgletContext().getProperty("font-
size", SIZE));

rgb=Integer.parseInt(_aglet.getAgletContext().getProperty("back-
ground-color", "0"));
          } catch (NumberFormatException e) {}

if (rgb == 0)
          setBackground(Color.lightGray);
      else
          setBackground(new Color(rgb));

      setFont(new Font(
          _aglet.getAgletContext().getProperty("window-font",
          FONT),
                  Font.PLAIN,
                  fs));

      try {  add("North", makeButtonPanel());
             add("Center", makeMainPanel());
      } catch (AgletException ae) {
           throw ae;}

       _helpWindow = new MessageDialog(this, "URL format error",
MAL_FORMED_URL_MSG);
      _helpWindow.setButtons(MessageDialog.OKAY);
```

```
        pack();
        resize(preferredSize());
        show();}

     private Panel makeTitlePanel() throws AgletException {
        Panel p = new Panel();
        p.setLayout(new FlowLayout(FlowLayout.LEFT));

        p.add(new Label(TITEL));
        return p;}

  private static int FIELD = 40;
     private TextField _URLString = new TextField(FIELD);
     private Choice _hotlist = new Choice();
     private TextArea _result = new TextArea();

     private Panel makeMainPanel() throws AgletException {
         initList(_hotlist, _aglet.getAgletContext().getProp-
erty("url-hotlist", ""));

        // set first item in the choice as the default ARL
        if (_hotlist.countItems() > 0) {
              _URLString.setText(_hotlist.getItem(0));}

        Panel p = new Panel();
        Panel p1 = new Panel();
        p1.setLayout(new GridLayout(2,1));
```

6.4.5 Class FingerSlave: slave aglet

Again, the first part of the source for this piece of the agent defines the Java classes that are going to be used and their interfaces. The class definitions are not included in the text but can be referenced on the CD-ROM.

```
  public class FingerSlave extends Slave {

     public synchronized void doJob() {
        RESULT = getLocalInfo();
        int i  = 2;

       // i = (i-1)/(i-2);  // debugging: simulate a run-time error
     }

     protected Object getLocalInfo() {
       AgletContext ac = null;
       try { ac=getAgletContext();
       } catch (AgletException ae) {
       return ae.getMessage();}

       String hostname = "Unknown";
```

```
        try { hostname=ac.getHostingURL().getHost().toString();
        } catch (AgletException ae) {
           //--
         }

         FingerInfo info =
               new FingerInfo (hostname,
                        ac.getProperty("user-name", "Unknown"),
                        ac.getProperty("org-name", "Unknown"),
                        ac.getProperty("email-address", "Unknown"),
                        ac.getProperty("registered", "Unknown"),
                        (new Date()));
         return info;}}
```

6.4.6 Class FingerInfo: data for exchange

Yet again, the first part of the source for this piece of the agent defines the Java classes that are going to be used and their interfaces. The class definitions are not included in the text but can be referenced on the CD-ROM.

```
    public final class FingerInfo {

       // Private variables
       String _hostName;
       String _userName;
       String _orgName;
       String _eMail;
       String _registered;
       Date _localTime;

       // Public methods
       public FingerInfo(String hostName,  String userName, String
orgName,
                  String eMail, String registered, Date localTime) {
       _hostName = hostName;
       _userName = userName;
       _orgName = orgName;
       _eMail = eMail;
       _registered = registered;
       _localTime = localTime;}

       public String getHostName() {
          return _hostName;}

       public String getUserName() {
          return _userName;}

       public String getOrgName() {
```

```
      return _orgName;}

   public Date getLocalTime() {
      return _localTime;}

   public String getRegistered () {
        return _registered;}

   public String getEmail() {
      return _eMail;}

   public String toTextBlock () {
     String str = "Host Name: " + _hostName + "\n" +
                  "User Name: " + _userName + "\n" +
                  "Organization Name: " + _orgName + "\n" +
                  "E-mail Address: " + _eMail + "\n"  +
                  "Registered: " + _registered + "\n" +
                  "Local Time: " + _localTime.toString();
     return str;}}
```

chapter 7

Aglets workbench packages

This chapter is a complement to the information provided in the previous chapter, Aglets Workbench. Because the aglets system is based on Java, there is not a comprehensive architecture for the system, as there is in the agent systems presented in Part II.

What is provided here is the complete set of Java packages, classes, interfaces, exceptions, and variables. This, along with the underpinnings of the Java environments, is the architecture of Aglets.

For more information on Java and how it relates to agent issues, a good reference is *Java Network Programming* by Merlin Hughes, et al. (Full reference can be found in the bibliography on page 300.)

7.1 Aglets packages

The hierarchy of the classes, interfaces, and exceptions that make up the aglets solution are presented here. This gives a high-level presentation of the classes and where they fit into the complete solution. The rest of this chapter provides the complete definitions of all of the classes, interfaces, and exceptions needed to build aglets agents.

aglets

```
Class Index
   Aglet
   AgletContext
   AgletCookie
   AgletIdentifier
   AgletInputStream
   AgletLoader
   AgletOutputStream
   AgletProxy
   Log
   ObservableEvent
   ThreadManager
   Version
Exception Index
   AgletException
```

aglets.patterns

```
Interface Index
   Master
   Receiver
Class Index
   Itinerary
```

```
Messenger
Notification
Notifier
Slave
Utils
```

aglets.util

```
Class Index
   ImageComponent
   MessageDialog
   MessagePanel
```

atp

```
Class Index
   AtpDaemon
   AtpDaemonConsole
   AtpInputStream
   AtpNullOutputStream
   AtpOutputStream
   AtpRequestHandler
   AtpURLConnection
   AtpURLStreamHandler
   AtpURLStreamHandlerFactory
Exception Index
   AtpException
```

atp.handler

```
Class Index
   AgletsAtpRequestHandler
```

7.2 Classes, methods, exceptions, and variables

What follows is the complete listing of the classes, methods, and exceptions shown in the hierarchy above, along with related variables and explanations of all related issues.

Class **aglets.Aglet**

```
public class Aglet
extends Object
```

The `Aglet` class is the abstract base class for aglets. Use this class to create your own personalized aglets.

Variables

VERSION

```
protected static Version VERSION
```

Subclasses can set their version number. The version of an aglet is retrieved by `getVersion()`.

See also:

getVersion()

Constructors

Aglet

```
protected Aglet()
```

Creates an uninitialized aglet. Normally you should not override this method. Instead if you need to initialize the aglet you should override `initialize()`.

See also:

`Object()`

Methods

cloneThis

```
public final synchronized AgletIdentifier cloneThis() throws Aglet
    Exception
```

Clones the aglet in the current execution context.

Throws:

`AgletException`

if the cloning fails.

clone

```
protected synchronized Object clone() throws CloneNotSupportedEx-
    ception
```

Clones the aglet. Normally, subclasses to aglet do not have to override the default method. The default method uses the object serialization mechanism to create the clone.

Returns:

the cloned aglet.

Throws:

`CloneNotSupportedException`

when the cloning fails.

Overrides:

`clone` in class `Object`

dispatchThis

```
public final synchronized AgletIdentifier dispatchThis(URL url)
     throws AgletException
```

Dispatches this aglet to the location specified by the argument `url`.

Parameters:

`url`—dispatch destination.

Returns:

the identity of the dispatched aglet.

Throws:

`AgletException`

if the dispatch operation fails.

disposeThis

```
public final synchronized void disposeThis() throws AgletException
```

Removes this aglet from the current execution context.

Throws:

`AgletException`

if the disposal fails.

getAgletContext

```
public final AgletContext getAgletContext() throws AgletException
```

Gets the execution context that the aglet is currently running in.

Returns:

the current execution context.

Throws:

`AgletException`

if the aglet has no context.

getIdentity

```
public final AgletIdentifier getIdentity() throws AgletException
```

Returns the identity of this aglet.

Throws:

`AgletException`

if the identity is undefined.

getProperty

```
public final String getProperty(String key)
```

Gets the aglet property indicated by the key.

Parameters:

`key`—the name of the aglet property.

Returns:

the value of the specified key.

getProperty

```
public final String getProperty(String key,
            String defValue)
```

Gets the aglet property indicated by the key and default value.

Parameters:

`key`—the name of the aglet property.

`defValue`—the default value to use if this property is not set.

Returns:

the value of the specified key.

initialize

```
protected synchronized void initialize(Object init) throws Aglet
    Exception
```

Initializes the aglet. This method is called once in the lifecycle of an aglet. An aglet can override this method to perform specific initializations. Subclasses that override this method should explicitly call the default implementation in the base class (`super.initialize(init)`). The `init` argument allows subclasses to utilize initialization data.

Parameters:

`init`—initialization data.

Throws:

`AgletException`

if the initialization fails.

onArrival

```
protected synchronized void onArrival() throws AgletException
```

Called when the aglet is arriving from a remote server. This method can be overridden to implement actions to be taken upon the arrival to a context.

Throws:

`AgletException`
if retraction is denied.

onDialog
```
protected synchronized void onDialog()
```

Requests the aglet to enter into a dialog with the user. This method should be overridden in subclasses to provide a dialog GUI.

onDispatching
```
protected synchronized void onDispatching(URL url) throws AgletEx-
    ception
```

Called when the aglet is dispatched. This method can be overridden in subclasses to implement actions to be taken upon a dispatch request. If you want to create an immobile aglet, override this method to throw an `AgletException`.

Parameters:

`url`—dispatch destination.

Throws:

`AgletException`
if the dispatch request is rejected.

onDisposal
```
protected synchronized void onDisposal() throws AgletException
```

Called when the aglet is disposed. This method can be overridden in subclasses to implement actions to be taken upon a disposal request.

Throws:

`AgletException`
if the request for disposal is rejected.

onRetraction
```
protected synchronized void onRetraction(URL remoteURL) throws
    AgletException
```

Called when the aglet is retracted from a remote server. This method can be overridden to implement actions to be taken upon a retract request. If you want to create an aglet that cannot be retracted from its current location, override this method to throw an `AgletException`.

Parameters:

`remoteURL`—source of retraction request.

Throws:

`AgletException`

if retraction is denied.

setMessage

```
protected final synchronized void setMessage(String message)
```

Sets the message line of this aglet. A way for the aglet to display messages without opening a window.

Parameters:

`message`—the message.

propertyKeys

```
public final Enumeration propertyKeys()
```

Enumerates all the property keys.

Returns:

property key enumeration.

run

```
protected void run()
```

This method is the entry point for the aglet's own thread. You should override this method to let the aglet actively perform some task after initialization.

Class aglets.AgletContext

```
public final class AgletContext
extends Observable
```

The `AgletContext` class is the execution context for running aglets. It provides means for maintaining and managing running aglets in an environment where the aglets are protected from each other and the host system is secured against malicious aglets.

Constructors

AgletContext

```
public AgletContext(Properties prop,
     Vector log,
     ThreadManager atm)
```

Creates an execution context for aglets.

Parameters:

`prop`—property list.

`log`—the execution log.

`atm`—the thread manager.

createAglet

```
public AgletProxy createAglet(URL url,
    String name,
    Object init) throws AgletException
```

Creates an instance of the specified aglet located at the specified URL.

Parameters:

`url`—the URL to load the aglet class from

`name`—the aglet's class name.

`init`—initialization data.

Returns:

a proxy for the newly instantiated and initialized aglet.

Throws:

`AgletException`

when the method failed to instantiate the aglet.

createAglet

```
public AgletProxy createAglet(URL url,
    String name,
    Object init,
  int mode) throws AgletException
```

Creates an instance of the specified aglet located at the specified URL.

Parameters:

`url`—the URL to load the aglet class from.

`name`—the aglet's class name.

`init`—initialization data.

`mode`—the execution mode of the aglet.

Returns:

a proxy for the newly instantiated and initialized aglet.

Throws:

`AgletException`

when the method failed to instantiate the aglet.

cloneAglet

```
public AgletProxy cloneAglet(Aglet aglet) throws AgletException
```

Clones the specified aglet.

Parameters:

`aglet`—the aglet to be cloned.

Returns:

the proxy for the cloned aglet.

Throws:

`AgletException`

if the clone method is not supported.

dispatchAglet

```
public AgletIdentifier dispatchAglet(Aglet aglet,
              URL url) throws AgletException
```

Dispatches the specified aglet to the specified destination.

Parameters:

`aglet`—the aglet to be dispatched.

`url`—the dispatch destination.

Returns:

the identity of the dispatched aglet.

Throws:

`AgletException`

when the method failed to dispatch the aglet.

disposeAglet

```
public void disposeAglet(Aglet aglet) throws AgletException
```

Disposes the specified aglet.

Parameters:

`aglet`—the aglet to be disposed.

Throws:

`AgletException`

if the aglet cannot be found.

getAgletProxies

```
public synchronized Enumeration getAgletProxies()
```

Gets the aglet proxies in the current execution context.

Returns:

an enumeration of aglet proxies.

getAglet

```
public AgletProxy getAglet(AgletIdentifier identity) throws Aglet
     Exception
```

Gets the proxy for an aglet specified by its identity.

Parameters:

`identity`—the identity of the aglet.

Returns:

the aglet proxy.

Throws:

`AgletException`

if the aglet is not found.

getHostingURL

```
public static URL getHostingURL() throws AgletException
```

Returns the URL of the daemon serving all current execution contexts.

Throws:

`AgletException`

if the hosting URL cannot be determined.

getProperty

```
public String getProperty(String key)
```

Gets the context property indicated by the key.

Parameters:

`key`—the name of the context property.

Returns:

the value of the specified key.

getProperty

```
public String getProperty(String key,
                String def)
```

Gets the context property indicated by the key and default value.

Parameters:

`key`—the name of the context property.

`def`—the value to use if this property is not set.

Returns:

the value of the specified key.

retractAglet

```
public AgletProxy retractAglet(URL url) throws AgletException
```

Retracts the Aglet specified by its url: `atp://host-domain-name/[user-name]#aglet-identity`.

Parameters:

`url`—the location and aglet identity of the aglet to be retracted.

Returns:

the aglet proxy for the retracted aglet.

Throws:

`AgletException`

when the method failed to retract the aglet.

`retractAglet`

```
public AgletProxy retractAglet(URL url,
              int mode) throws AgletException
```

Retracts the aglet specified by its url.

Parameters:

`url`—the location and aglet identity of the aglet to be retracted.

`mode`—the execution mode

Returns:

the aglet proxy for the retracted aglet.

Throws:

`AgletException`

when the method failed to retract the aglet.

`receiveAglet`

```
public AgletProxy receiveAglet(Aglet aglet) throws AgletException
```

Receives an aglet. Will start the aglet and return its proxy.

Parameters:

`aglet`—the aglet to be received by the context.

Throws:

`AgletException`

if it is not received.

`receiveAglet`

```
public AgletProxy receiveAglet(Aglet aglet,
              int mode) throws AgletException
```

Receives an aglet. Will start the aglet and return its proxy.

Parameters:

`aglet`—the aglet to be received by the context.

`mode`—the execution mode of the received aglet.

Throws:

`AgletException`

if it is not received.

revertAglet

```
public Aglet revertAglet(URL remoteURL,
              AgletIdentifier identity) throws AgletException
```

setHostingURL

```
public static void setHostingURL(URL url) throws AgletException
```

Sets the URL of the daemon serving all current execution contexts.

Throws:

`AgletException`

if the hosting URL already has been set.

propertyKeys

```
public Enumeration propertyKeys()
```

Enumerates all the property keys.

Returns:

property key enumeration.

Class **aglets.AgletCookie**

```
public final class AgletCookie
extends Object
```

The `AgletCookie` class provides a cookie mechanism to restrict access to aglets.

Variables

NO_COOKIE

```
public static AgletCookie NO_COOKIE
```

The static `NoKey` method generates an authentication-less key.

Constructors

AgletCookie

```
public AgletCookie()
```

Creates an aglet key with a randomly generated password.

AgletCookie

```
public AgletCookie(String passwd) throws AgletException
```

Creates an aglet key with the specified password. The password should only consist of letter (upper and lower case), digits, and space characters.

Parameters:

`passwd`—the password.

Throws:

`AgletException`

if the password argument is invalid.

Methods

`equals`

```
public boolean equals(Object cookie)
```

Compares two aglet cookies.

Parameters:

`cookie`—the cookie to be compared with.

Returns:

true if and only if the two cookies are identical.

Overrides:

equals in class Object

Class `aglets.AgletException`

```
public class AgletException
extends Exception
```

Signals that an aglet exception has occurred.

Constructors

`AgletException`

```
public AgletException()
```

Constructs an `AgletException` with do detail message. A detail message is a string that describes this particular exception.

`AgletException`

```
public AgletException(String s)
```

Constructs an `AgletException` with the specified detail message. A detail message is a string that describes this particular exception.

Parameters:

`s`—the detail message.

Class `aglets.AgletIdentifier`

```
public final class AgletIdentifier
extends Object
```

The `AgletIdentifier` class implements a naming mechanism for aglets. An aglet identifier consists of two components: the family identity and the dispatch identity. The family identity is set upon creation of the aglet. Clones of a given aglet share the same family identity but are given unique dispatch identities.

Variables

DEFAULT
`public static AgletIdentifier DEFAULT`

The default aglet identity.

NULL
`public static AgletIdentifier NULL`

The null aglet identity.

Constructors

AgletIdentifier
`public AgletIdentifier(String textForm)`

Creates an aglet identifier from an unparsed text representation: `familyIdentity:dispatchIdentity`.

Parameters:
`textForm`—the unparsed text representation of the Alget identity.

Methods

clone
`protected Object clone() throws CloneNotSupportedException`

Creates an Aglet identifier with same family identity and unique dispatch identity.

Returns:
the cloned Aglet identifier.

Throws:
`CloneNotSupportedException`
is actually never thrown.

Overrides:
`clone` in class `Object`

equals

`public boolean equals(Object obj)`

Compares two aglet identifiers.

Parameters:

`obj`—the Aglet to be compared with.

Returns:

true if and only if the two Aglets are identical.

Overrides:

`equals` in class `Object`

getDispatchId

`public String getDispatchId()`

Gets the dispatch identity.

Returns:

the dispatch identity in text form.

getFamilyId

`public String getFamilyId()`

Gets the family identity.

Returns:

the family identity in text form.

hashCode

`public int hashCode()`

Returns an integer suitable for hash table indexing.

Returns:

hash table indexing integer.

Overrides:

`hashCode` in class `Object`

isNull

`public boolean isNull()`

Tells whether this identifier is a null identifier.

toString

`public String toString()`

Returns a human readable form of the aglet identifier.

Returns:

the Aglet identity in text form.

Overrides:

`toString` in class `Object`

Class **aglets.AgletInputStream**

```
public class AgletInputStream
extends ObjectInputStream
```

An instance of this class reads objects from an input stream which contains class data with objects. The input stream contains objects, class data of these objects and class data of all super classes of these classes. Data in the input stream must be written by an instance of the `AgletOutputStream`.

This aglet input stream looks into the class loader cache of the `AgletLoader` and gets a class loader corresponding to the URL of an origin of the received class. If the class loader is not found in the cache, this stream will creates a new class loader and put it into the cache. After getting the class loader, this stream gets a class of the received object from the class cache of the class loader. If the class is not found in the class cache, the class loader will create the class.

See also: `AgletOutputStream`, `AgletLoader`

Constructors

AgletInputStream

```
public AgletInputStream(InputStream in) throws IOException,
  StreamCorruptedException
```

Creates a new instance of this class.

Parameters:

`in`—an input stream containing objects and class data.

Throws:

`IOException`

if cannot read data from the input stream.

Throws:

`StreamCorruptedException`

if data in the input stream is invalid.

Methods

resolveClass

```
protected Class resolveClass(String classname) throws IOException,
  ClassNotFoundException
```

Resolves a class specified by `classname`. This method reads class data from the input stream and resolves the class by using a class loader corresponding to the origin of the class. If the class is common class, the class will be resolved by `super.resolveClass`. This method reads class data of all super classes of the class in the input stream and puts them into the class data cache of the class loader. These super classes will be resolved on demand.

Parameters:

`classname`—class name.

Returns:

the resolved class.

Throws:

`IOException`

if cannot read data from the input stream.

Throws:

`ClassNotFoundException`

if cannot resolve the class.

Overrides:

`resolveClass` in class `ObjectInputStream`

Class **aglets.AgletLoader**

```
public class AgletLoader
extends ClassLoader
```

Class `AgletLoader` is the abstract base class for Aglet loaders. This class loader has a class loader cache as a static member and stores all class loader objects with keys. The keys are the URL of the origin of classes and are managed by the loader. In aglets, bytecodes of classes are transferred with objects. Therefore, there may be many versions of classes whose names are the same on an aglet server. In aglets, classes are managed based on their origin. When the source of numerous classes is the same, the bytecodes are managed by the same aglet loader in an aglet server. Therefore, an object can access objects if the sources of their class bytecodes are the same, otherwise `ClassCastException` will occur.

An aglet loader caches classes and their bytecodes.

Variable Index

`_classCache`
Cache to store classes.

`_classDataCache`
Cache to store bytecodes of classes.

Constructor Index

`AgletLoader()`

Method Index

`finalize()`
Shouts when an `AgletLoader` object is caught by GC.

`getByteCodeFromCache(String)`
Gets a bytecode of a desired class from the class data cache.

`getClassFromCache(String)`
Gets a desired class from the class cache.

`instantiateAglet(URL, String)`
Instantiates the `Aglet` class specified by name.

`instantiateObject(URL, String)`
Instantiates the `Object` class specified by name.

`loadClass(String, boolean)`
Loads a class specified by the param name.

`putByteCodeIntoCache(String, byte[])`
Puts a bytecode of a class into the class data cache.

`putClassIntoCache(String, Class)`
Puts a class into the class cache.

`removeByteCodeFromCache(String)`
Removes a bytecode from the class data cache.

Variables

`_classCache`

`protected Hashtable _classCache`

Caches to store classes. All classes loaded by an `AgletLoader` object are stored into it.

_classDataCache
`protected Hashtable _classDataCache`

Caches to store bytecodes of classes. All bytecodes of classes loaded by an `AgletLoader` object are stored into it.

Constructors

AgletLoader
`public AgletLoader()`

Methods

finalize
`protected void finalize()`

Shout when an `AgletLoader` object is caught by GC. This method is for verifying whether a class loader becomes a target of GC or not.

Overrides:
`finalize` in class `Object`

getByteCodeFromCache
`protected synchronized byte[] getByteCodeFromCache(String name)`

Gets a bytecode of a desired class from the class data cache.

Parameters:
`name`—the name of the desired class.
Returns:
the bytecode of the class. If the class is not found, null will be returned.

getClassFromCache
`protected synchronized Class getClassFromCache(String name)`

Gets a desired class from the class cache.

Parameters:
`name`—the name of the desired class.
Returns:
the class. If the class is not found null will be returned.

instantiateAglet
```
protected static Aglet instantiateAglet(URL url,
                String name) throws AgletException
```

Instantiates the `Aglet` class specified by name. Load a class from the specified URL.

Parameters:

`url`—the URL. If the url is null, default aglet loader is used. The loader searches `CLASSPATH` and gets a bytecode of the class.

`name`—the class name.

Returns:

the Aglet.

Throws:

`AgletException`

is thrown when the method failed to instantiate the Aglet.

instantiateObject

```
protected static Object instantiateObject(URL url,
                String name) throws AgletException
```

Instantiates the `Object` class specified by name. Load a class from the specified URL.

Parameters:

`url`—the URL. If the url is null, default aglet loader is used. The loader searches `CLASSPATH` and gets a bytecode of the class.

`name`—the class name.

Returns:

the Aglet.

Throws:

`AgletException`

is thrown when the method failed to instantiate the Aglet.

loadClass

```
protected Class loadClass(String name,
                boolean resolve) throws ClassNotFoundException
```

Loads a class specified by the param name. If a bytecode of the class has not been loaded, an `AgletLoader` object will load the bytecode and define the class. The loaded bytecode and class will be stored into the class data cache and the class cache respectively.

Parameters:

`name`—the name of the desired class.

`resolve`—true if the class must be resolved.

Returns:

a loaded class

Throws:

`ClassNotFoundException`

if the class is not found.

Overrides:

`loadClass` in class ClassLoader.

putClassIntoCache

```
protected synchronized void putClassIntoCache(String name,
            Class cl)
```

Puts a class into the class cache.

Parameters:

`name`—the class name.

`cl`—the class.

putByteCodeIntoCache

```
protected synchronized void putByteCodeIntoCache(String name,
            byte bytecode[])
```

Puts a bytecode of a class into the class data cache.

Parameters:

`name`—the name of the class.

`bytecode`—the bytecode of the class.

removeByteCodeFromCache

```
protected synchronized void removeByteCodeFromCache(String name)
```

Removes a bytecode from the class data cache.

Parameters:

`name`—the name of the class.

Class **aglets.AgletOutputStream**

```
public class AgletOutputStream
extends ObjectOutputStream
```

An instance of this class writes objects and class data into an output stream. This aglet output stream writes objects, class data of these objects and class data of all super classes of these classes. Data written into the output stream must be read by an instance of the `AgletInputStream`. This aglet output stream writes the name of the class, the URL of its origin, length of class data and class data. If the class is common whose package is `java`, `atp` or `aglets`, the class data will not be written.

See also: `AgletInputStream`

Variable Index

`COMMON`
Indicates that a class written into this stream is common.

Constructor Index

`AgletOutputStream(OutputStream)`
Creates a new instance of this class.

Method Index

`annotateClass(Class)`
Writes the class data into the output stream.

Variables

`COMMON`
`public final static String COMMON`

Indicates that a class written into this stream is common.

Constructors

`AgletOutputStream`
`public AgletOutputStream(OutputStream out) throws IOException`

Creates a new instance of this class.

Parameters:
`out`—an output stream where data are written into.

Throws:
`IOException`
if cannot write into the output stream.

`annotateClass`
`protected void annotateClass(Class cl) throws IOException`

Writes the class data into the output stream. Class data of all super classes of the class will be written together.

Parameters:
`cl`—class.

Throws:

`IOException`

if cannot write into the output stream.

Overrides:

`annotateClass` in class `ObjectOutputStream`

Class **aglets.AgletProxy**

```
public final class AgletProxy
extends Object
```

Class `AgletProxy` is a placeholder for aglets. The purpose of this class is to provide a mechanism to control and limit direct access to aglets.

Method Index

`dialog()`
Requests a dialog with the aglet.

`getAglet(AgletCookie)`
Gets the aglet.

`getArrival()`
Gets the time of proxy creation.

`getClassName()`
Gets the aglet's class name.

`getCodeBase()`
Gets the URL of the aglet's class.

`getIdentity()`
Gets the aglet's identity.

`getMessage()`
Gets the current content of the Aglet's message line.

`getProperty(String)`
Gets the aglet property indicated by the key.

`getProperty(String, String)`
Gets the aglet property indicated by the key and default value.

`propertyKeys()`
Enumerates all the property keys.

Methods

getAglet

`public Aglet getAglet(AgletCookie cookie) throws AgletException`

Gets the aglet. If the aglet is access protected, it will require the right key to get access.

Parameters:

`key`—the aglet access key.

Returns:

the aglet.

Throws:

`AgletException`

if the cookie is missing.

See also: `AgletKey`

getIdentity

`public AgletIdentifier getIdentity() throws AgletException`

Gets the aglet's identity.

Returns:

the aglet's identity.

Throws:

`AgletException`

if the identity is undefined.

dialog

`public void dialog()`

Requests a dialog with the aglet. This method call will be forwarded to the `onDialog` in the aglet.

See also: `onDialog`

getClassName

`public String getClassName()`

Gets the aglet's class name.

Returns:

the class name.

getCodeBase

`public URL getCodeBase()`

Gets the URL of the aglet's class. Null is returned if the class is in the set of common classes.

Returns: the class `URL`.

getArrival

```
public Date getArrival()
```

Gets the time of proxy creation. This reflects the time since the aglet last resumed execution in the current context.

getMessage

```
public String getMessage()
```

Gets the current content of the Aglet's message line.

Returns:
the message line.

getProperty

```
public final String getProperty(String key)
```

Gets the aglet property indicated by the key.

Parameters:
`key`—the name of the aglet property.
Returns:
the value of the specified key.

getProperty

```
public final String getProperty(String key,
                String defValue)
```

Gets the aglet property indicated by the key and default value.

Parameters:
`key`—the name of the aglet property.
`defValue`—the default value to use if this property is not set.
Returns:
the value of the specified key.

propertyKeys

```
public final Enumeration propertyKeys()
```

Enumerates all the property keys.

Returns:
property key enumeration.

Class `aglets.Log`

```
public class Log
extends Object
```

`Class Log` is used to log aglet events such as `CREATE`, `CLONE`, `DISPOSE`, `DISPATCH`, `RETRACT`, `RECEIVE`, and `REVERT`.

Variable Index

```
CLONE
CREATE
DISPATCH
DISPOSE
ERROR
MESSAGE
RECEIVE
RETRACT
REVERT
UNDEFINED
```

Constructor Index

```
Log(int, Aglet)
Log(int, Aglet, URL)
Log(int, String)
```

Method Index

```
atpUrlText()
getAgletURL()
getDate()
getIdentity()
getMessage()
getType()
getURL()
toMessageText()
toString()
urlText()
```

Variables

`UNDEFINED`

public final static int UNDEFINED

`CREATE`

```
public final static int CREATE
```

CLONE
public final static int CLONE

DISPOSE
public final static int DISPOSE

DISPATCH
public final static int DISPATCH

RETRACT
public final static int RETRACT

RECEIVE
public final static int RECEIVE

REVERT
public final static int REVERT

ERROR
public final static int ERROR

MESSAGE
public final static int MESSAGE

Constructors

Log
public Log(int tp,
Aglet aglet)

Log
public Log(int tp,
Aglet aglet,
URL arl)

Log
public Log(int tp,
String message)

Methods

getType
public int getType()

getDate
public Date getDate()

getIdentity
public AgletIdentifier getIdentity()

getURL
public URL getURL()

urlText
`public String urlText()`

getAgletURL
`public URL getAgletURL()`

atpUrlText
`public String atpUrlText()`

getMessage
`public String getMessage()`

toMessageText
`public String toMessageText()`

toString
`public String toString()`

Overrides:

`toString` in class `Object`

Class **aglets.ObservableEvent**

```
public class ObservableEvent
extends Object Class
```

Variable Index

```
id
INSERT
MESSAGE
NONE
REMOVE
target
UPDATE
```

Constructor Index

```
ObservableEvent(int)
ObservableEvent(Object, int)
```

Variables

NONE
`public final static int NONE`

INSERT
`public final static int INSERT`

REMOVE

`public final static int REMOVE`

UPDATE

`public final static int UPDATE`

MESSAGE

`public final static int MESSAGE`

id

`public int id`

target

`public Object target`

Constructors

ObservableEvent

`public ObservableEvent(int id)`

ObservableEvent

`public ObservableEvent(Object target, int id)`

Class **aglets.patterns.Itinerary**

```
public class Itinerary
extends Object
```

The `Itinerary` class enables the aglet's developer to define and take control over an aglet's tour. Specifically:

1. To use an `Itinerary` object for a specific aglet, the `setTraveler` method should be initially applied with this specific aglet as its argument.
2. An `Itinerary` object can be bound to at most one specific aglet.
3. By default, when an aglet is dispatched, its original copy is disposed.

Defined vocabulary:

tour: a set of destinations to visit.

place: a destination within a tour.

origin: the place which is considered as the original place from which the tour is started.

handler: the place to which an aglet can be dispatched in case of unexpected events. The origin is the default handler.

log: a record of messages to trace the aglet's tour.

Variable Index

`_traveler`

Constructor Index

`Itinerary(URL)`
Creates an `Itinerary` with a single destination.

`Itinerary(URL, Vector)`
Creates an `Itinerary` with a single destination.

`Itinerary(Vector)`
Creates an `Itinerary`.

Method Index

`clearLog()`
Clears the log (remove all entries).

`dispatchHandler(URL)`
Dispatches an aglet to a specific URL.

`getCurrentURL()`
Tells the URL of the current visited place on the tour.

`gotoByIndex(int)`
Dispatches an aglet to a new place with a specific index.

`gotoErrorHandler()`
Dispatches an aglet to the handler.

`gotoNext()`
Dispatches an aglet to the next available place in the tour.

`gotoOrigin()`
Dispatches an aglet to the origin.

`isAtHandler()`
Tells whether the aglet is currently visiting the `Handler` host.

`isAtLastDestination()`
Tells whether the aglet is currently visiting the last place in the tour.

`isAtOrigin()`
Tells whether the aglet is currently located in the `Origin` host

`logToString()`

Converts the log to a string.

`reset()`

Resets an aglet's tour as if the tour have not yet been started.

`setDestruction(boolean)`

Defines whether to dispose of an aglet after it is dispatched or not.

`setHandler(URL)`

Defines the URL of a host to which the aglet can be dispatched in cases of unexpected events during its tour.

`setNumAllowedFailures(int)`

Defines maximum number of allowed failures to visit a place during a tour.

`setNumRetries(int)`

Defines maximum number of retries to dispatch an aglet to a new place.

`setTraveller(Aglet)`

Binds an aglet (for the tour).

Variables

`_traveler`

```
protected Aglet _traveler
```

Constructors

`Itinerary`

```
public Itinerary(URL destination)
```

Creates an `Itinerary` with a single destination.

Parameters:

`destination`—the `URL` of the destination to be visited.

`Itinerary`

```
public Itinerary(Vector itinVector)
```

Creates an `Itinerary`.

Parameters:

`itinVector`—a `vector` of URLs to visit.

`Itinerary`

```
public Itinerary(URL origin,
                 Vector itinVector)
```

Creates an `Itinerary` with a single destination.

Parameters:

`itinVector`—a vector of URLs to visit.

Methods

`gotoNext`
`public void gotoNext() throws AgletException`

Dispatches an aglet to the next available place in the tour.

`gotoByIndex`
`public void gotoByIndex(int index) throws AgletException`

Dispatches an aglet to a new place with a specific index.

`reset`
`public void reset() throws AgletException`

Resets an aglet's tour as if the tour have not yet been started. The aglet is dispatched to the origin host.

`gotoOrigin`
`public void gotoOrigin() throws AgletException`

Dispatches an aglet to the origin.

`gotoErrorHandler`
`public void gotoErrorHandler() throws AgletException`

Dispatches an aglet to the handler.

`logToString`
`public String logToString()`

Converts the log to a string.

`clearLog`
`public void clearLog()`

Clears the log (remove all entries).

`setTraveler`
`public void setTraveller(Aglet aglet) throws AgletException`

Binds an aglet (for the tour).

Parameters:

`aglet`—the aglet to be bound.

setDestruction

`public void setDestruction(boolean b)`

Defines whether to dispose an aglet after it is dispatched or not.

setHandler

`public void setHandler(URL handler)`

Defines the `URL` of a host to which the aglet can be dispatched in cases of unexpected events during its tour.

setNumRetries

`public void setNumRetries(int num)`

Defines maximum number of retries to dispatch an aglet to a new place.

setNumAllowedFailures

`public void setNumAllowedFailures(int num)`

Defines maximum number of allowed failures to visit a place during a tour.

getCurrentURL

`public URL getCurrentURL()`

Tells the `URL` of the current visited place on the tour.

isAtOrigin

`public boolean isAtOrigin()`

Tells whether the aglet is currently located in the `Origin` host

isAtHandler

`public boolean isAtHandler()`

Tells whether the aglet is currently visiting the Handler host.

isAtLastDestination

`public boolean isAtLastDestination()`

Tells whether the aglet is currently visiting the last place in the tour.

dispatchHandler

`protected void dispatchHandler(URL url) throws AgletException`

Dispatches an aglet to a specific `URL`.

Throws:

`AgletException`

if no traveler exists or dispatching is failed.

`Interface aglets.patterns.Master`

```
public interface Master
extends Object
```

This interface is a part of the master-slave usage pattern. The role of the master is to create slaves that get dispatched to remote aglet servers. Master defines to method interfaces that incoming slaves use to either deliver their results or to report a state of error.

See also: Slave

Method Index

`callback(Slave, Object)`

Called when a slave has finished its job and wants to update its master with the result.

`inError(Slave, Object)`

Called when a slave catches an exception.

Methods

`callback`

```
public abstract void callback(Slave s,
                 Object result)
```

Called when a slave has finished its job and wants to update its master with the result.

Parameters:

`s`—the slave.

`result`—the result of the slave's job.

`inError`

```
public abstract void inError(Slave s,
                Object message)
```

Called when a slave catches an exception. If the exception happens at a remote site, it will attempt to return and call this method in the master.

Parameters:

`s`—the slave.

`message`—the slave's error message.

Class **aglets.patterns.Messenger**

```
public final class Messenger
extends Aglet
```

The `Messenger` class is a part of the messenger usage pattern. Create a messenger by calling the static method create. The messenger will get dispatched automatically.

```
Messenger.create(getAgletContext(), someURL, message);
```

See Also: Receiver

Constructor Index

`Messenger()`

Method Index

`create(AgletContext, URL, Object)`
Creates a messenger.

`initialize(Object)`
Initializes the messenger.

`run()`
Universal entry point for the messenger's execution thread.

Constructors

`Messenger`
`public Messenger()`

Methods

`create`

```
public static AgletProxy create(AgletContext context,
            URL destination,
              Object message) throws AgletException
```

Creates a messenger.

Parameters:

`context`—the aglet context in which the messenger should be created.

`destination`—the destination `URL` for the messenger.

`message`—the message object.

Returns:

an aglet proxy for the messenger.

Throws:

`AgletException`
if initialization fails.

`initialize`

```
protected synchronized void initialize(Object obj) throws Agle
     tException
```

Initializes the messenger. The argument object is a vector with two elements. The first element is the destination URL and the second element is the message object.

Throws:

`AgletException`
if initialization fails.

Overrides:

initialize in `class Aglet`

`run`

```
public void run()
```

Universal entry point for the messenger's execution thread.

Overrides:

run in `class Aglet`

Class `aglets.patterns.Notification`

```
public final class Notification
extends Object
```

The `Notification` class is a part of the `Notify` usage pattern. An instance of the `Notification` class is used for transferring information.

See also: Notify

Variable Index

```
EXCEPTION
EXPIRY
NOTIFICATION
```

Constructor Index

`Notification(int, URL, Date, int, Object)`
Constructs the `Notification` to be sent.

Method Index

```
getKind()
getMessage()
getNotifier()
getNumber()
getTime()
```

Variables

NOTIFICATION
```
public final static int NOTIFICATION
```

EXPIRY
```
public final static int EXPIRY
```

EXCEPTION
```
public final static int EXCEPTION
```

Constructors

Notification
```
public Notification(int kind,
 URL notifier,
 Date time,
 int number,
 Object message)
```

Constructs the `Notification` to be sent.

Parameters:

`kind`—specify norman notification, expiration, or error exception
`notifier`—is `URL` for identifying the sender aglet.
`number`—is used when `Notifier` stays at remote server to send multiple instances of `Notification`. It is used for numbering the message.
`message`—an argument Object; contents of the message.

Methods

getKind
```
public int getKind()
```

getNotifier
```
public URL getNotifier()
```

getTime
```
public Date getTime()
```

getNumber
`public int getNumber()`

getMessage
`public Object getMessage()`

Class **aglets.patterns.Notifier**

```
public class Notifier
extends Aglet
```

The `Notifier` class is a part of the Notifier usage pattern.
See also: Notification

Variable Index

`ARGUMENT`

The protected variable that carries any arguments for the checks that this notifier performs.

`MESSAGE`

The protected variable that carries any messages that should go along with the notification back to the subscriber.

Constructor Index

`Notifier()`

Method Index

`create(URL, String, AgletContext, Aglet, Itinerary, double, double, boolean, Object)`
Creates a notifier.

`doCheck(Object)`
This method should be overridden to specify the check method for this notifier.

`getReceiver()`
Gets the URL of this notifier's receiver.

`initialize(Object)`
Initializes the notifier.

`initializeCheck(Object)`

This method should be overridden to specify the intial check performed by this notifier.

`run()`

Universal entry point for the notifier's execution thread.

Variables

MESSAGE

`protected Object MESSAGE`

The protected variable that carries any messages that should go along with the notification back to the subscriber. This object should be persistent.

ARGUMENT

`protected Object ARGUMENT`

The protected variable that carries any arguments for the checks that this notifier performs. This object should be persistent.

Constructors

Notifier

`public Notifier()`

Methods

initializeCheck

`public abstract void initializeCheck(Object arg)`

This method should be overridden to specify the intial check performed by this notifier.

Returns:

the result of the initial check.

doCheck

`public abstract boolean doCheck(Object arg)`

This method should be overridden to specify the check method for this notifier.

Returns:

the result of the initial check.

getReceiver

`protected URL getReceiver()`

Gets the URL of this notifier's receiver.

create

```
public static AgletProxy create(URL url,
String source,
AgletContext context,
Aglet receiver,
Itinerary destination,
double interval,
double duration,
boolean stay,
Object argument) throws AgletException
```

Creates a notifier.

Parameters:

`url`—the location of the aglet class placed.

`source`—the aglet class definition.

`context`—the aglet context in which the messenger should be created.

`receiver`—the URL of the receiver.

`destination`—the URL of the destination.

`interval`—the time in hours between two checks.

`duration`—the life time of the notifier.

`stay`—whether the notifier should remain after a notification.

`argument`—an argument object.

Returns:

an aglet proxy for the notifier.

Throws:

`AgletException`

if initialization fails.

initialize

```
protected synchronized void initialize(Object obj) throws AgletEx-
    ception
```

Initializes the notifier. Only called the very first time this notifier is created. The argument for a notifier is a `NotifierSettings` object.

Parameters:

`obj`—the `NotifierSettings` object.

Throws:

`AgletException`

if the initialization fails.

Overrides:

initialize in `class Aglet`

```
run
public void run()
```

Universal entry point for the notifier's execution thread.

Overrides:

run in `class Aglet`

Interface **aglets.patterns.Receiver**

```
public interface Receiver
extends Object
```

This interface is a part of the `Messenger` usage pattern. It specifies the method that messengers will call in order to deliver a message. Classes inheriting this interface will have to implement the message method.

See also: Messenger

Method Index

`message(Messenger, Object)`

Called when a messenger arrives with a message.

Methods

message

```
public abstract void message(Messenger m,
   Object mes)
```

Called when a messenger arrives with a message.

Parameters:

`m`—the messenger.

`mes`—the message.

Class **aglets.patterns.Slave**

```
public class Slave
extends Aglet
```

The `Slave` class is a part of the master-slave usage pattern. Create a slave by calling the static method `create`. The slave will get dispatched automatically.

```
Slave.create(getAgletContext(), master, itinerary, argument);
```

The itinerary is a vector of one or more ARLs:

```
Vector theItinerary = new Vector();
theItinerary.addElement(firstARL);
theItinerary.addElement(secondARL);
theItinerary.addElement(thirdARL);
```

See also: Master

Variable Index

`ARGUMENT`
The protected that carries any arguments for the task that this slave performs.

`RESULT`
The protected variable that keeps the result of task that this slave performs.

Constructor Index

`Slave()`

Method Index

`create(URL, String, AgletContext, Aglet, Itinerary, Object)`
Creates a slave.

`create(URL, String, AgletContext, Aglet, Vector, Object)`
Creates a slave.

`doJob()`
This method should be overridden to specify the job of the slave.

`getMaster()`
Gets the URL of this slave's master.

`initialize(Object)`
Initializes the slave.

`run()`
Universal entry point for the slave's execution thread.

Variables

`RESULT`

```
protected Object RESULT
```

The protected variable that keeps the result of task that this slave performs. This object should bepersistent.

ARGUMENT
`protected Object ARGUMENT`

The `protected object` that carries any arguments for the task that this slave performs. This object should be persistent.

Constructors

Slave
`public Slave()`

Methods

doJob
`public abstract void doJob()`

This method should be overridden to specify the job of the slave.

Returns:
the result of the job as an Object

getMaster
`protected URL getMaster()`

Gets the `URL` of this slave's master.

create
```
public static AgletProxy create(URL url,
   String name,
   AgletContext context,
   Aglet master,
   Itinerary itinerary,
   Object argument) throws AgletException
```

Creates a slave.

Parameters:
`source`—the aglet source.
`context`—the aglet context in which the slave should be created.
`master`—the URL of the master.
`itinerary`—destinations defined as an Itinerary object.
`argument`—an `argument` object.

Returns:
an aglet proxy for the slave.

Throws:

`AgletException`
if initialization fails.

create

```
public static AgletProxy create(URL url,
         String name,
         AgletContext context,
           Aglet master,
           Vector itinerary,
             Object argument) throws AgletException
```

Creates a slave.

Parameters:

`source`—the aglet source.
`context`—the aglet context in which the slave should be created.
`master`—the URL of the master.
`itinerary`—a vector of destinations.
`argument`—an argument object.

Returns:

an aglet proxy for the slave.

Throws:

`AgletException`
if initialization fails.

initialize

```
protected synchronized void initialize(Object obj) throws AgletEx-
     ception
```

Initializes the slave. Only called the very first time this slave is created. The argument for a slave is a vector with four elements: the first is the URL of the master, the second element is a a destination vector containing one or more URLs, the third argument is the initial value of the protected result variable, and the fourth element is the initial value of protected argument variable.

Parameters:

`obj`—the argument vector.

Throws:

`AgletException`
if the initialization fails.

Overrides:

initialize in `class Aglet`

run
`public void run()`

Universal entry point for the slave's execution thread.

Overrides:
run in class Aglet

Class **aglets.patterns.Utils**

```
public final class Utils
extends Object
```

Constructor Index

`Utils()`

Constructors

Utils
`public Utils()`

Class **aglets.ThreadManager**

```
public class ThreadManager
extends Object
```

The `ThreadManager` class is the base class that implements the basic thread policy for aglets. An instance of this class (or a user-defined subclass is given as argument to the aglet context for it to enact the given thread policy.

Constructor Index

`ThreadManager()`
Constructs the thread manager.

Method Index

`checkMode(int)`
Compares the requested execution mode to the current mode.

`getAgletGroup()`
Gets the current aglet thread group.

`getAgletIdentity()`
Gets the aglet identity of the current aglet thread group.

`getCurrentMode()`
Gets the mode of the current aglet thread group.

`getDefaultMode()`
Returns the default execution mode.

`makeThreadGroup(Aglet, int)`
Creates the thread group that the aglet's thread is put into.

Constructors

`ThreadManager`
`public ThreadManager()`

Constructs the thread manager.

Methods

`getCurrentMode`
`public int getCurrentMode()`

Gets the mode of the current aglet thread group.

Returns:
the execution mode.

`getDefaultMode`
`public int getDefaultMode()`

Returns the default execution mode.

`checkMode`
`public int checkMode(int mode)`

Compares the requested execution mode to the current mode.

Returns:
the resulting execution mode.

`getAgletIdentity`
`public AgletIdentifier getAgletIdentity()`

Gets the aglet identity of the current aglet thread group.

Returns:
the aglet identity.

`getAgletGroup`
`protected AgletThreadGroup getAgletGroup()`

Gets the current aglet thread group.

Returns:

the aglet thread group.

makeThreadGroup

```
protected ThreadGroup makeThreadGroup(Aglet aglet,
                int mode) throws AgletException
```

Creates the thread group that the aglet's thread is put into. This method should be overridden in a sub-class to implement a specific aglet thread policy.

Parameters:

`aglet`—the aglet.

`mode`—the execution for the aglet.

Returns:

the new aglet thread group.

Throws:

`AgletException`

if it fails to create a new thread group.

Class **aglets.util.ImageComponent**

```
public class ImageComponent
extends Canvas
```

Constructor Index

```
ImageComponent(Image)
ImageComponent(Image, Dimension)
```

Method Index

```
bounds()
imageUpdate(Image, int, int, int, int, int)
inside(int, int)
move(int, int)
paint(Graphics)
preferredSize()
printStringCenter(String, Graphics)
setMessage(String)
size()
```

Constructors

ImageComponent

```
public ImageComponent(Image image)
```

ImageComponent

```
public ImageComponent(Image image,
              Dimension image_size)
```

Methods

setMessage

```
public void setMessage(String message)
```

paint

```
public void paint(Graphics g)
```

Overrides:

paint in `class Canvas`

printStringCenter

```
public void printStringCenter(String str,
              Graphics g)
```

imageUpdate

```
public boolean imageUpdate(Image img,
         int flags,
         int x,
         int y,
         int w,
         int h)
```

Overrides:

`imageUpdate` in `class Component`

bounds

```
public Rectangle bounds()
```

Overrides:

`bounds` in `class Component`

size

```
public Dimension size()
```

Overrides:

`size` in `class Component`

move

```
public void move(int x, int y)
```

Overrides:

move in class Component

inside

`public boolean inside(int x, int y)`

Overrides:

`inside` in `class Component`

preferredSize

`public Dimension preferredSize()`

Overrides:

`preferredSize` in `class Component`

Class **aglets.util.MessageDialog**

```
public class MessageDialog
extends Dialog
```

The `MessageDialog` class is a common and generic dialog to display the messages.

Variable Index

```
ALWAYS_CENTER
APPLY
CANCEL
CENTER_ONLY_ONCE
FREE
HELP
OKAY
```

Constructor Index

```
MessageDialog(Frame, Component, String, String, int, Object,
     boolean)
MessageDialog(Frame, String, String)
MessageDialog(Frame, String, String, Object)
```

Method Index

`addButton(Button)`

Adds extra button you like.

`beep()`

Rings a bell.

`getButton(int)`
Obtains dialog button.

`handleEvent(Event)`
Handles the events.

`hide()`
`pack()`
`popup(Frame)`
Pops up the dialog window so that it get located at the center of the frame.

`popup(int)`
Pops up the dialog window according to the location given as a parameter

`setButtons(int)`
Specifies which buttons should be appeared on the bottom of window.

`setCallbackComponent(Component)`
`setMessage(String)`
Sets the message

`waitForDisposal()`
Waits until the dialog window is disposed.

Variables

OKAY
`public final static int OKAY`

CANCEL
`public final static int CANCEL`

APPLY
`public final static int APPLY`

HELP
`public final static int HELP`

ALWAYS_CENTER
`public final static int ALWAYS_CENTER`

CENTER_ONLY_ONCE
`public final static int CENTER_ONLY_ONCE`

FREE
`public final static int FREE`

Constructors

MessageDialog

```
public MessageDialog(Frame parent,
        String title,
        String message)
```

MessageDialog

```
public MessageDialog(Frame parent,
        String title,
        String message,
Object object)
```

MessageDialog

```
public MessageDialog(Frame parent,
        Component callback_component,
        String title,
        String message,
        int alignment,
        Object object,
        boolean modal)
```

Methods

setCallbackComponent

```
protected void setCallbackComponent(Component c)
```

popup

```
public void popup(Frame frame)
```

Pops up the dialog window so that it get located at the center of the frame. If frame is null, parent frame will be used.

setMessage

```
public void setMessage(String msg)
```

Sets the message.

popup

```
public void popup(int location)
```

Pops up the dialog window according to the location given as a parameter.

handleEvent

```
public boolean handleEvent(Event event)
```

Handles the events. Any `BUTTON PRESS` event will be delegated to the callback component.

Overrides:

`handleEvent` in `class Component`

hide
`public synchronized void hide()`

Overrides:

`hide` in `class Component`

waitForDisposal
`public synchronized void waitForDisposal() throws InterruptedException`

Waits until the dialog window is disposed. The window system thread cannot be used for this purpose. Aglet thread can be used instead.

beep
`public void beep()`

Rings a bell. This is a tentative method and should move to more common class for aglets.

setButtons
`public void setButtons(int b)`

Specifies which buttons should appear on the bottom of window.

Parameters:

`b`—logical `OR` value of constants, `OKAY`, `CANCEL`, `HELP`, `APPLY`.

addButton
`public void addButton(Button b)`

Adds extra button you like.

Parameters:

`b`—the button to get added.

pack
`public void pack()`

Overrides:

`pack` in `class Window`

getButton
`public Button getButton(int b)`

Obtains dialog button.

Class **aglets.util.MessagePanel**

```
public class MessagePanel
extends Panel
```

The `MessagePanel` class is a common and generic dialog to display the messages.

Constructor Index

```
MessagePanel(String, boolean)
MessagePanel(String, int, boolean)
```

Method Index

```
paint(Graphics)
```

Constructors

MessagePanel

```
public MessagePanel(String message,
              boolean raised)
```

MessagePanel

```
public MessagePanel(String message,
         int alignment,
              boolean raised)
```

Methods

paint

```
public void paint(Graphics g)
```

Overrides:

`paint` in `class Component`

Class **aglets.Version**

```
public final class Version
extends Object
```

Class `Version` is used to create version objects that contain release information, such as `MAJOR`, `MINOR`, `BUILD`, `DATE`, and `KIND`. Version objects can also be created with an expiration date. Any attempt to create an instance of the `version` object after the expiration date will fail and result in a call to `System.exit()`. Creating the `version` object in the period 0 to 15 days before expiration will result in a warning message written by `System.out.println()`. Values of a `version` object cannot be altered once it has been created.

Variable Index

`ALPHA`
Alpha version.

`BETA`
Beta version.

Constructor Index

`Version(String, int, int, int, Date)`
Creates a version object which will never expire.

Method Index

`getBuild()`
Gets the build number x.x.B.

`getDate()`

Gets the date of the version.

`getExpiration()`
Gets the date of expiration.

`getKind()`
Gets the string that describes what is versioned.

`getLongText()`
Returns a long text representation of the version numbers: for example, Alpha2b, Beta1, V2.0.

`getMajor()`
Gets the major version number M.x.x.

`getMajorText()`
Returns a text representation of the major version number: for example, Alpha, Beta, V2.

`getMinor()`
Gets the minor version number x.M.x.

`getShortText()`

Returns a short text representation of the version numbers: e.g., A2b, B1, V2.0.

`toString()`

Returns a human readable form of the `version` object.

Variables

ALPHA

```
public final static int ALPHA
```

Alpha version.

BETA

```
public final static int BETA
```

Beta version.

Constructors

Version

```
public Version(String kind,
        int major,
        int minor,
        int build,
        Date date)
```

Creates a version object which will never expire.

Parameters:

`kind`—product information.
`major`—major version number M.x.x
`minor`—minor version number x.M.x
`build`—build version number x.x.B
`date`—date of this version.

Methods

getKind

```
public String getKind()
```

Gets the string that describes what is versioned.

getMajor

```
public int getMajor()
```

Gets the major version number M.x.x.

getMinor
`public int getMinor()`

Gets the minor version number x.M.x.

getBuild
`public int getBuild()`

Gets the build number x.x.B.

getDate
`public Date getDate()`

Gets the date of the version.

getExpiration
`public Date getExpiration()`

Gets the date of expiration. Returns null if expiration date is not set.

getShortText
`public String getShortText()`

Returns a short text representation of the version numbers: for example, A2b, B1, V2.0.

getLongText
`public String getLongText()`

Returns a long text representation of the version numbers: for example, Alpha2b, Beta1, V2.0.

getMajorText
`public String getMajorText()`

Returns a text representation of the major version number: for example, Alpha, Beta, V2.

toString
`public String toString()`

Returns a human readable form of the `version` object.

Overrides:
`toString` in `class Object`

Class atp.AtpDaemon

```
public class AtpDaemon
extends Object
implements Runnable
```

Constructor Index

`AtpDaemon(String[])`
This constructor will parse the command line options and then start the daemon.

Method Index

`exit()`
Terminate this.

`getHostName()`
Gets the host domain name of the machine that is currently hosting the aglet listener.

`getPort()`
Gets the port number that the aglet listener is currently listening to.

`getProperty(String, String)`
Gets property.

`main(String[])`
The main program of the ATP daemon.

`restart()`
Called to restart the aglet listener on the same port.

`restart(int)`
Called to restart the aglet listener on a new port.

`run()`
The entry point for the daemon thread.

`setProperty(String, String)`
Sets property of this.

`start()`
Called to start the aglet listener.

`stop()`
Called to stop the aglet listener.

Constructors

AtpDaemon

```
public AtpDaemon(String args[])
```

This constructor will parse the command line options and then start the daemon.

Parameters:

`args`—command line options.

Methods

main
```
public static void main(String args[])
```

The main program of the ATP daemon.

setProperty
```
public synchronized void setProperty(String key,
              String value)
```

Set property of this.

Parameters:

`key`—a key string for the property.

`value`—value of the property.

getProperty
```
public synchronized String getProperty(String key,
             String default_value)
```

Gets property.

Parameters:

`key`—a key string for the property.

`default_value`—value if the key is not found.

Returns:

value of the property.

start
```
public void start()
```

Called to start the aglet listener.

stop
```
public synchronized void stop()
```

Called to stop the aglet listener.

restart
```
public synchronized void restart()
```

Called to restart the aglet listener on the same port.

restart
```
public synchronized void restart(int p)
```

Called to restart the aglet listener on a new port.

Parameters:

p—the new port number.

exit

`public void exit()`

Terminate this.

getHostName

`public String getHostName()`

Gets the host domain name of the machine that is currently hosting the aglet listener.

Returns:

the host name.

getPort

`public int getPort()`

Gets the port number that the aglet listener is currently listening to.

Returns:

the port number.

run

`public void run()`

The entry point for the daemon thread.

Class **atp.AtpDaemonConsole**

```
public class AtpDaemonConsole
extends Object
implements Runnable
```

Variable Index

_daemon

_localHost

_thread

CRLF

DEF_CONF_DIR

DEF_CONF_FILE

Constructor Index

`AtpDaemonConsole()`

Method Index

`close()`
Close this.

`error(String, String)`
Writes an error message string.

`errorString(String, String)`
Creates an error message string.

`errorWrite(String)`
Writes an error message into the output stream.

`getLine(FileInputStream)`
Gets a line from the configuration file.

`initDefConfigFile(FileOutputStream)`
Writes initial strings in the configuration file.

`initialize(AtpDaemon, String)`
Initializes an instance of this class.

`log(String, Date, Date, AtpInputStream)`
Writes a log message.

`logString(String, Date, Date, AtpInputStream)`
Creates a log message string.

`logWrite(String)`
Writes a log message into the output stream.

`message(String)`
Writes a message into the output stream.

`messageString(String)`
Creates a message string.

`messageWrite(String)`
Write a message into the output stream.

`resolveConfigFile(String)`
Reads the config file and get parameters.

`run()`

This method is called from the console thread.

`setProperty(String)`

Sets properties to the atp daemon.

`start()`

Starts this.

Variables

CRLF

`protected final static String CRLF`

DEF_CONF_FILE

`protected final static String DEF_CONF_FILE`

DEF_CONF_DIR

`protected final static String DEF_CONF_DIR`

_daemon

`protected AtpDaemon _daemon`

_localHost

`protected String _localHost`

_thread

`protected Thread _thread`

Constructors

AtpDaemonConsole

`public AtpDaemonConsole()`

Methods

initialize

```
public void initialize(AtpDaemon daemon,
                String configFile) throws AtpException
```

Initializes an instance of this class.

Parameters:

`daemon`—atp daemon.

`configFile`—configuration filename. If this parameter is null, create a default configuration file in the user's home directory.

Throws:

`AtpException`

if cannot initialize this.

initDefConfigFile

`protected void initDefConfigFile(FileOutputStream out) throws IOException`

Writes initial strings in the configuration file. This file is created when there is no configuration file.

Parameters:

`out`—a file output stream. Don't close this.

Throws:

`IOException`

if cannot write.

resolveConfigFile

`public void resolveConfigFile(String confFile) throws AtpException`

Reads the config file and gets parameters.

Parameters:

`confFile`—configuration file name.

Throws:

`AtpException`

if cannot read the file.

getLine

`protected String getLine(FileInputStream in) throws IOException`

Gets a line from the configuration file.

Parameters:

`in`—a file input stream of the file.

Returns:

a string of the line.

Throws:

`IOException`

if cannot read the file.

setProperty

`protected void setProperty(String line)`

Sets properties to the atp daemon.

Parameters:

`line`—string of the line containing parameters.

close
```
public void close()
```

Closes this. This method will be called when this is closed.

start
```
public void start()
```

Starts this.

run
```
public abstract void run()
```

This method is called from the console thread.

log
```
public synchronized void log(String remoteHost,
     Date in_time,
     Date out_time,
     AtpInputStream in) throws IOException
```

Writes a log message.

Parameters:

`remoteHost`—a host name of a sender.

`in_time`—time when the `AtpDaemon` received the request message.

`out_time`—time when the `AtpDaemon` finished to handle the request message.

in—an atp input stream containing the request message.

Throws:

`IOException`

if cannot write a log message.

error
```
public synchronized void error(String remoteHost,
            String str) throws IOException
```

Writes an error message string.

Parameters:

`remoteHost`—a host name of a sender.

`err`—error string.

Returns:

error message.

message
```
public synchronized void message(String str) throws IOException
```

Writes a message into the output stream.

Parameters:

`message`—message string.

Throws: `IOException`

if cannot write the message into the output stream.

logString
```
protected String logString(String remoteHost,
        Date in_time,
        Date out_time,
               AtpInputStream in)
```

Creates a log message string.

Parameters:

`remoteHost`—a host name of a sender.

`in_time`—time when the AtpDaemon received the request message.

`out_time`—time when the AtpDaemon finished to handle the request message.

`in`—an atp input stream containing the request message.

Returns:

log message.

logWrite
```
protected abstract void logWrite(String log) throws IOException
```

Writes a log message into the output stream.

Parameters:

`log`—log message.

Throws:

`IOException`

if cannot write the log message into the output stream.

errorString
```
protected String errorString(String remoteHost,
             String err)
```

Creates an error message string.

Parameters:

`remoteHost`—a host name of a sender.

`err`—error string.

Returns:

error message.

errorWrite

`protected abstract void errorWrite(String error) throws IOException`

Writes an error message into the output stream.

Parameters:

`error`—error message.

messageString

`protected String messageString(String str)`

Creates a message string.

Parameters:

`str`—message.

Returns:

message string.

messageWrite

`protected abstract void messageWrite(String message) throws IOException`

Writes a message into the output stream.

Parameters:

`message`—message string.

Throws:

`IOException`

if cannot write the message into the output stream.

Class **atp.AtpException**

```
public class AtpException
extends Exception
```

Signals that an Atp exception has occurred.

Constructor Index

`AtpException()`

Constructs an `AtpException` with do detail message.

`AtpException(String)`

Constructs an `AtpException` with the specified detail message.

Constructors

`AtpException`

`public AtpException()`

Constructs an `AtpException` with do detail message. A detail message is a string that describes this particular exception.

`AtpException`

`public AtpException(String s)`

Constructs an `AtpException` with the specified detail message. A detail message is a string that describes this particular exception.

Parameters:

`s`—the detail message.

Class **`atp.AtpInputStream`**

```
public class AtpInputStream
extends InputStream
```

The atp input stream receives a message of ATP protocol from the specified input stream and resolve it.

Variable Index

`_in`
An input stream containing ATP messages.

`ATP_REQUEST`
The value representing that an ATP message is a request.

`ATP_RESPONSE`
The value representing that an ATP message is a response.

`DISPATCH`
The method token for `DISPATCH`.

`FETCH`
The method token for `FETCH`.

`RETRACT`
The method token for `RETRACT`.

Constructor Index

`AtpInputStream(InputStream)`
Creates a new instance of AtpInputStream.

Method Index

`getBody()`
Gets an input stream containing a body of the read data.

`getHeaderField(String)`
Gets a value of the header field read from this input stream with specified key.

`getHeaderField(String, String)`
Gets a value of the header field read from this input stream with specified key.

`getMessageType()`
Gets message type read from this input stream.

`getMethod()`
Gets method token read from this input stream.

`getReasonPhrase()`
Gets a reason phrase read from this input stream.

`getStatusCode()`
Gets a status code read from this input stream.

`getURI()`
Gets URI from this input stream.

`read()`
Reads a next byte of data from this input stream.

Variables

`ATP_REQUEST`
`public final static int ATP_REQUEST`
The value representing that an ATP message is a request.

`ATP_RESPONSE`
`public final static int ATP_RESPONSE`
The value representing that an ATP message is a response.

`DISPATCH`
`public final static String DISPATCH`
The method token for `DISPATCH`.

RETRACT

`public final static String RETRACT`

The method token for `RETRACT`.

FETCH

`public final static String FETCH`

The method token for `FETCH`.

_in

`protected InputStream _in`

An input stream containing ATP messages. An atp input stream reads ATP messages from it.

Constructors

AtpInputStream

`public AtpInputStream(InputStream is) throws IOException`

Creates a new instance of `AtpInputStream`.

Parameters:

`is`—an instance of `InputStream` from which the instantiated atp input stream reads.

Methods

read

`public int read() throws IOException`

Reads a next byte of data from this input stream.

Returns:

the next byte of data.

Throws:

`IOException`

if cannot read from this stream.

Overrides:

`read` in class `InputStream`

getMessageType

`public int getMessageType()`

Gets message type read from this input stream. The message type is either `ATP_REQUEST` or `ATP_RESPONSE`.

Returns:

message type.

getMethod

```
public String getMethod() throws AtpException
```

Gets method token read from this input stream.

Returns:

method token.

Throws:

`AtpException`

if the read message is a response message.

getURI

```
public String getURI() throws AtpException
```

Gets URI from this input stream.

Returns:

a string representing URI. `:exception AtpException` if the read message is a response message.

getStatusCode

```
public int getStatusCode() throws AtpException
```

Gets a status code read from this input stream.

Returns:

a status code.

Throws: `AtpException`

if the read message is a request message.

getReasonPhrase

```
public String getReasonPhrase() throws AtpException
```

Gets a reason phrase read from this input stream.

Returns:

a reason phrase.

Throws: `AtpException`

if the read message is a request message.

getBody

```
public InputStream getBody()
```

Gets an input stream containing a body of the read data.

Returns:

an input stream.

getHeaderField

```
public String getHeaderField(String key,
              String default_value)
```

Gets a value of the header field read from this input stream with specified key.

Parameters:

`key`—a key.

`default_value`—a default value.

Returns:

the value read from this input stream with specified key value. If the key is not found, return the default value.

getHeaderField

```
public String getHeaderField(String key)
```

Gets a value of the header field read from this input stream with specified key.

Parameters:

`key`—a key.

Returns:

the value read from this stream with specified key value. If the key is not found, return null.

Class **atp.AtpNullOutputStream**

```
public class AtpNullOutputStream
extends OutputStream
```

Constructor Index

`AtpNullOutputStream(PrintStream)`

Method Index

`close()`

Closes the stream.

`flush()`

Flushes the stream.

`write(byte[])`
`write(byte[], int, int)`
`write(int)`

Constructors

AtpNullOutputStream

```
public AtpNullOutputStream(PrintStream out)
```

Methods

write

```
public void write(int b) throws IOException
```

Parameters:

`b`—the byte

Throws:

`IOException`

if an I/O error has occurred.

Overrides:

`write` in `class OutputStream.`

write

```
public void write(byte b[]) throws IOException
```

Parameters:

`b`—the data to be written

Throws:

`IOException`

if an I/O error has occurred.

Overrides:

`write` in `class OutputStream`

write

```
public void write(byte b[],
      int off,
      int len) throws IOException
```

Parameters:

`b`—the data to be written

`off`—the start offset in the data

`len`—the number of bytes that are written

Throws: `IOException`

if an I/O error has occurred.

Overrides:

`write` in `class OutputStream`

flush

```
public void flush() throws IOException
```

Flushes the stream. This will write any buffered output bytes.

Throws:

`IOException`

if an I/O error has occurred.

Overrides:

`flush` in class `OutputStream`

`close`

`public void close() throws IOException`

Closes the stream. This method must be called to release any resources associated with the stream.

Throws:

`IOException`

if an I/O error has occurred.

Overrides:

`close` in class `OutputStream`

Class **`atp.AtpOutputStream`**

```
public class AtpOutputStream
extends OutputStream
```

The atp output stream writes a message of ATP protocol to the specified output stream. The atp output stream creates header fields of ATP protocol, combines it with a message body, and writes it into the specific output stream.

Variable Index

`_out`

An output stream into which ATP messages are written.

`AGENT_LANGUAGE`

A key value for `AGENT-LANGUAGE` field in the message's header.

`AGENT_SYSTEM`

A key value for `AGENT-SYSTEM` field in the message's header.

`ATP_VERSION`

ATP version.

BAD_GATEWAY

Status code representing that the recipient, while acting as a gateway or proxy, received an invalid response from upstream server.

BAD_REQUEST

Status code representing that the recipient was unable to understand the request message due to malformed syntax.

CONTENT_ENCODING

A key value for CONTENT-ENCODING field in the message's header.

CONTENT_LENGTH

A key value for CONTENT-LENGTH field in the message's header.

CONTENT_TYPE

A key value for CONTENT-TYPE field in the message's header.

COOKIE

A key value for COOKIE field in the message's field.

CRLF

A separator in the message's header.

DATE

A key value for DATE field in the message's header.

FORBIDDEN

Status code representing that although the recipient understood the request message, it refused to fulfill it.

FROM

A key value for FROM field in the message's header.

INTERNAL_ERROR

Status code representing that the recipient encountered an unexpected condition which prevented it from fulfilling the request.

MOVED

Status code representing that the requested resource is no longer at the recipient.

NOT_FOUND

Status code representing that the recipient could not find the requested resource.

NOT_IMPLEMENTED

Status code representing that the recipient does not support the functionality required to fulfill the request.

OKAY

Status code representing that the request has succeeded.

RECIPIENT_APPLICATION

A key value for RECIPIENT-APPLICATION field in the message's header.

RP_BAD_GATEWAY

Reason phrase representing that the recipient, while acting as a gateway or proxy, received an invalid response from upstream server.

RP_BAD_REQUEST

Reason phrase representing that the recipient was unable to understand the request message due to malformed syntax.

RP_FORBIDDEN

Reason phrase representing that although the recipient understood the request message, it refused to fulfill it.

RP_INTERNAL_ERROR

Reason phrase representing that the recipient encountered an unexpected condition which prevented it from fulfilling the request.

RP_MOVED

Reason phrase representing that the requested resource is no longer at the recipient.

RP_NOT_FOUND

Reason phrase representing that the recipient could not find the requested resource.

RP_NOT_IMPLEMENTED

Reason phrase representing that the recipient does not support the functionality required to fulfill the request.

RP_OKAY

Reason phrase representing that the request has succeeded.

RP_SERVICE_UNAVAILABLE

Reason phrase representing that the recipient is currently unable to handle the request due to a temporary overloading of the recipient.

SENDER_APPLICATION

A key value for SENDER-APPLICATION field in the message's header.

SERVICE_UNAVAILABLE

Status code representing that the recipient is currently unable to handle the request due to a temporary overloading of the recipient.

Constructor Index

`AtpOutputStream(OutputStream)`
Create a new instance of `AtpOutputStream`.

Method Index

`dispatch(byte[])`
Writes a dispatch request message with a body into this stream.

`fetch(String)`
Writes a fetch request message into this stream.

`getHeaderField(String)`
Gets a value of a header field specified by a key.

`getHeaderField(String, String)`
Gets a value of a header field specified by a key.

`respond(int)`
Writes a respond message into this stream.

`respond(int, byte[])`
Writes a respond message into this stream.

`retract(String)`
Writes a retract request message into this stream.

`retract(String, String)`
Writes a retract request message into this stream.

`setHeaderField(String, String)`
Sets a value of a header field specified by a key.

`write(int)`
Writes a byte into this stream.

Variables

_out
`protected OutputStream _out`
An output stream into which ATP messages is written. An atp output stream writes ATP messages into it.

AGENT_LANGUAGE
`public final static String AGENT_LANGUAGE`
A key value for `AGENT-LANGUAGE` field in the message's header.

AGENT_SYSTEM

`public final static String AGENT_SYSTEM`

A key value for `AGENT-SYSTEM` field in the message's header.

ATP_VERSION

`public final static String ATP_VERSION`

ATP version.

BAD_GATEWAY

`public final static int BAD_GATEWAY`

Status code representing that the recipient, while acting as a gateway or proxy, received an invalid response from upstream server.

BAD_REQUEST

`public final static int BAD_REQUEST`

Status code representing that the recipient was unable to understand the request message due to malformed syntax.

CONTENT_TYPE

`public final static String CONTENT_TYPE`

A key value for `CONTENT-TYPE` field in the message's header.

CONTENT_LENGTH

`public final static String CONTENT_LENGTH`

A key value for `CONTENT-LENGTH` field in the message's header.

CONTENT_ENCODING

`public final static String CONTENT_ENCODING`

A key value for `CONTENT-ENCODING` field in the message's header.

COOKIE

`public final static String COOKIE`

A key value for `COOKIE` field in the message's field.

CRLF

`protected final static String CRLF`

A separator in the message's header.

DATE

`public final static String DATE`

A key value for `DATE` field in the message's header.

FORBIDDEN

public final static int `FORBIDDEN`

Status code representing that the recipient understood the request message, and refuses to fulfill it.

FROM

`public final static String FROM`

A key value for `FROM` field in the message's header.

INTERNAL_ERROR

`public final static int INTERNAL_ERROR`

Status code representing that the recipient encountered an unexpected condition which prevented it from fulfilling the request.

MOVED

`public final static int MOVED`

Status code representing that the requested resource is no longer at the recipient.

NOT_FOUND

`public final static int NOT_FOUND`

Status code representing that the recipient could not find the requested resource.

NOT_IMPLEMENTED

`public final static int NOT_IMPLEMENTED`

Status code representing that the recipient does not support the functionality required to fulfill the request.

OKAY

`public final static int OKAY`

Status code representing that the request has succeeded.

SERVICE_UNAVAILABLE

`public final static int SERVICE_UNAVAILABLE`

Status code representing that the recipient is currently unable to handle the request due to a temporary overloading of the recipient.

RP_BAD_GATEWAY

`public final static String RP_BAD_GATEWAY`

Reason phrase representing that the recipient, while acting as a gateway or proxy, received an invalid response from upstream server.

RP_BAD_REQUEST

`public final static String RP_BAD_REQUEST`

Reason phrase representing that the recipient was unable to understand the request message due to malformed syntax.

RP_FORBIDDEN

`public final static String RP_FORBIDDEN`

Reason phrase representing that although the recipient understood the request message, it refused to fulfill it.

RP_INTERNAL_ERROR

`public final static String RP_INTERNAL_ERROR`

Reason phrase representing that the recipient encountered an unexpected condition which prevented it from fulfilling the request.

RP_MOVED

`public final static String RP_MOVED`

Reason phrase representing that the requested resource is no longer at the recipient.

RP_NOT_FOUND

`public final static String RP_NOT_FOUND`

Reason phrase representing that the recipient could not find the requested resource.

RP_NOT_IMPLEMENTED

`public final static String RP_NOT_IMPLEMENTED`

Reason phrase representing that the recipient does not support the functionality required to fulfill the request.

RP_OKAY

`public final static String RP_OKAY`

Reason phrase representing that the request has succeeded.

RP_SERVICE_UNAVAILABLE

`public final static String RP_SERVICE_UNAVAILABLE`

Reason phrase representing that the recipient is currently unable to handle the request due to a temporary overloading of the recipient.

SENDER_APPLICATION

`public final static String SENDER_APPLICATION`

A key value for `SENDER-APPLICATION` field in the message's header.

RECIPIENT_APPLICATION

`public final static String RECIPIENT_APPLICATION`

A key value for `RECIPIENT-APPLICATION` field in the message's header.

Constructors

AtpOutputStream

`public AtpOutputStream(OutputStream os)`

Creates a new instance of `AtpOutputStream`.

Parameters:

`os`—an instance of `OutputStream` into which the instantiated atp output stream writes.

Methods

dispatch

`public void dispatch(byte obj[]) throws IOException`

Writes a dispatch request message with a body into this stream.

Parameters:

`obj`—a body.

Throws:

`IOException`

if cannot write the message into this stream.

fetch

`public void fetch(String path) throws AtpException, IOException`

Writes a `fetch` request message into this stream.

getHeaderField

```
public String getHeaderField(String key,
       String defaultvalue)
```

Gets a value of a header field specified by a key.

Parameters:

`key`—a key value.

`defaultvalue`—a default value.

Returns:

the value specified by a key value. If the key is not found, return the default value.

getHeaderField

`public String getHeaderField(String key)`

Gets a value of a header field specified by a key.

Parameters:

`key`—a key value.

Returns:

the value specified by a key value. If the key is not found, return null.

write

`public void write(int c) throws IOException`

Writes a byte into this stream.

Parameters:

`c`—a byte written into this stream.

Throws:

`IOException`

if cannot write a byte into this stream.

Overrides:

`write` in class `OutputStream`

respond

```
public void respond(int statusCode,
        byte body[]) throws IOException
```

Writes a respond message into this stream.

Parameters:

`statusCode`—a status code.

`body`—a body

Throws:

`IOException`

if cannot write the message into this stream.

respond

```
public void respond(int statusCode) throws IOException
```

Writes a respond message into this stream.

Parameters:

`statusCode`—a status code

Throws: `IOException`

if cannot write the message into this stream.

retract

```
public void retract(String aid) throws IOException
```

Writes a retract request message into this stream.

Parameters:

`aid`—an identifier for a retracted agent.

Throws: `IOException`

if cannot write the message into this stream.

retract

```
public void retract(String user,
        String aid) throws IOException
```

Writes a retract request message into this stream.

Parameters:

`user`—an identifier for an user of a retracted agent.

`aid`—an identifier for a retracted agent.

Throws:

`IOException`

if cannot write the message into this stream.

setHeaderField

```
public void setHeaderField(String key,
      String value)
```

Sets a value of a header field specified by a key.

Parameters:

`key`—a key value.

`value`—a value.

Class `atp.AtpRequestHandler`

```
public class AtpRequestHandler
extends Object
```

This class is an abstract class which handles atp request messages. `AtpDaemon` calls a request handler when the daemon receives an atp request message.

Constructor Index

`AtpRequestHandler()`

Creates a new instance of this class.

Method Index

`handleRequest(AtpInputStream, AtpOutputStream)`

Handles a request message.

`initialize(URL)`

Constructors

AtpRequestHandler

```
public AtpRequestHandler()
```

Creates a new instance of this class.

Methods

handleRequest

```
public abstract void handleRequest(AtpInputStream request,
       AtpOutputStream response)throws IOException,AtpException
```

Handles a request message. This method is an abstract method.

Parameters:

`request`—an atp input stream containing a request message.

`response`—an atp output stream to write a response message.

Throws:

`IOException`

if an atp input stream or an atp output stream has a problem.

Throws:

`AtpException`

if some exception occurs in this handler.

`initialize`

```
public abstract void initialize(URL hosting)
```

Class `atp.AtpURLConnection`

```
public class AtpURLConnection
extends URLConnection
```

An instance of this class creates a communication link between an application and an atp server.

Variable Index

`ATP_DEFAULT_PORT`

Default port number to receive atp messages.

Constructor Index

`AtpURLConnection(URL)`

Creates a new instance of this class.

Method Index

`connect()`

Makes a communication link with the destination.

`disconnect()`

Disconnects a communication link.

`getInputStream()`

Gets an input stream of the communication link.

`getOutputStream()`
Get an output stream of the communication link.

Variables

`ATP_DEFAULT_PORT`

`public final static int ATP_DEFAULT_PORT`

Default port number to receive atp messages.

Constructors

`AtpURLConnection`

`public AtpURLConnection(URL url)`

Creates a new instance of this class.

Parameters:

`url`—a destination URL to which the application connects. The protocol is atp.

Methods

`connect`

`public void connect() throws IOException`

Makes a communication link with the destination. If this object fails to make the communication link, it will retry several times with some interval. If it fails after several retries, it will throw an exception.

Throws:

`IOException`

if cannot make a communication link.

Overrides:

`connect` in `class URLConnection`

`disconnect`

`public void disconnect() throws IOException`

Disconnects a communication link.

Throws:

`IOException`

if disconnecting process finished with some error.

`getInputStream`

`public InputStream getInputStream() throws IOException`

Gets an input stream of the communication link.

Returns:

an input stream.

Throws:

`IOException`

if the communication link has a problem.

Overrides:

`getInputStream` in `class URLConnection`

`getOutputStream`

`public OutputStream getOutputStream() throws IOException`

Gets an output stream of the communication link.

Returns:

an output stream.

Throws:

`IOException`

if the communication link has a problem.

Overrides:

`getOutputStream` in `class URLConnection`

Class **`atp.AtpURLStreamHandler`**

```
public class AtpURLStreamHandler
extends URLStreamHandler
```

A stream protocol handler for atp protocol.

Constructor Index

`AtpURLStreamHandler()`

Method Index

`openConnection(URL)`

Opens a connection to the object referenced by the URL argument.

Constructors

`AtpURLStreamHandler`

`public AtpURLStreamHandler()`

Methods

openConnection

`public URLConnection openConnection(URL url)`

Opens a connection to the object referenced by the URL argument.

Parameters:

`url`—the URL that this connect to.

Returns:

an `AtpURLConnection` object for the URL.

Overrides:

`openConnection` in `class URLStreamHandler`

Class **atp.AtpURLStreamHandlerFactory**

```
public class AtpURLStreamHandlerFactory
extends Object
implements URLStreamHandlerFactory
```

This class is used by the URL class to create a `URLStreamHandler` for atp protocol.

Constructor Index

`AtpURLStreamHandlerFactory()`

Method Index

`createURLStreamHandler(String)`

Creates an instance of the `AtpURLStreamHandler` class.

`enableAtp()`

Sets an instance of this class to the stream handler collection of the `URL` class.

Constructors

AtpURLStreamHandlerFactory

`public AtpURLStreamHandlerFactory()`

Methods

createURLStreamHandler

`public URLStreamHandler createURLStreamHandler(String protocol)`

Creates an instance of the `AtpURLStreamHandler` class.

Returns:

an instance of the `AtpURLStreamHandler` class.

enableAtp

```
public static void enableAtp()
```

Sets an instance of this class to the stream handler collection of the `URL` class.

Class **atp.handler.AgletsAtpRequestHandler**

```
public class AgletsAtpRequestHandler
extends AtpRequestHandler
```

Constructor Index

```
AgletsAtpRequestHandler()
```

Method Index

```
handleRequest(AtpInputStream, AtpOutputStream)
initialize(URL)
```

Constructors

AgletsAtpRequestHandler

```
public AgletsAtpRequestHandler()
```

Methods

handleRequest

```
public void handleRequest(AtpInputStream request,
        AtpOutputStream response) throws IOException, AtpException
```

Overrides:

`handleRequest` in `class AtpRequestHandler`.

initialize

```
public void initialize(URL hosting)
```

Overrides:

`initialize` in `class AtpRequestHandler`.

index of all Aglets fields and methods

This chapter provides a complete, alphabetical listing of all of the fields and methods for the Aglets Workbench agent system. You can look through this list to quickly find the part of Aglets that you need to build and run your agent. With the information that you find here, you can further use the index to discover other places these features are covered in the book.

< A

`_classCache` Variable in class `aglets.AgletLoader`

Cache to store classes.

`_classDataCache` Variable in class `aglets.AgletLoader`

Cache to store bytecodes of classes.

`_daemon` Variable in class `atp.AtpDaemonConsole`

`_in` Variable in class `atp.AtpInputStream`

An input stream containing ATP messages.

`_localHost` Variable in class `atp.AtpDaemonConsole`

`_out` Variable in class `atp.AtpOutputStream`

An output stream into which ATP messages are written.

`_thread` Variable in class `atp.AtpDaemonConsole`

`_traveller` Variable in class `aglets.patterns.Itinerary`

A

addButton(Button) Method in class aglets.util.MessageDialog
Adds extra button you like.

AGENT_LANGUAGE Static variable in class atp.AtpOutputStream
A key value for AGENT_LANGUAGE field in the message's header.

AGENT_SYSTEM Static variable in class atp.AtpOutputStream
A key value for AGENT_SYSEM field in the message's header.

Aglet() Constructor for class aglets.Aglet
Creates an uninitialized aglet.

AgletContext(Properties, Vector, ThreadManager) Constructor for class aglets.AgletContext
Creates an execution context for aglets.

AgletCookie() Constructor for class aglets.AgletCookie
Creates an Aglet key with a randomly generated password.

AgletCookie(String) Constructor for class aglets.AgletCookie
Creates an Aglet key with the specified password.

AgletException() Constructor for class aglets.AgletException
Constructs an AgletException with do detail message.

AgletException(String) Constructor for class aglets.AgletException
Constructs an AgletException with the specified detail message.

AgletIdentifier(String) Constructor for class aglets.AgletIdentifier
Creates an aglet identifier from an unparsed text representation: familyIdentity:dispatchIdentity.

AgletInputStream(InputStream) Constructor for class aglets.AgletInputStream
Creates a new instance of this class.

AgletLoader() Constructor for class aglets.AgletLoader

AgletOutputStream(OutputStream) Constructor for class aglets.AgletOutputStream
Creates a new instance of this class.

AgletsAtpRequestHandler() Constructor for class atp.handler.AgletsAtpRequestHandler

ALPHA Static variable in class aglets.Version
Alpha version.

ALWAYS_CENTER Static variable in class `aglets.util.MessageDialog`

annotateClass(Class) Method in class `aglets.AgletOutputStream`
Writes the class data into the output stream.

APPLY Static variable in class `aglets.util.MessageDialog`

ARGUMENT Variable in class `aglets.patterns.Notifier`
The protected variable that carries any arguments for the checks that this notifier performs.

ARGUMENT Variable in class `aglets.patterns.Slave`
The protected variable that carries any arguments for the task that this slave performs.

ATP_DEFAULT_PORT Static variable in class `atp.AtpURLConnection`
Defaults port number to receive atp messages.

ATP_REQUEST Static variable in class `atp.AtpInputStream`
The value representing that an ATP message is a request.

ATP_RESPONSE Static variable in class `atp.AtpInputStream`
The value representing that an ATP message is a response.

ATP_VERSION Static variable in class `atp.AtpOutputStream`
ATP version.

AtpDaemon(String[]) Constructor for class `atp.AtpDaemon`
This constructor will parse the command line options and then start the daemon.

AtpDaemonConsole() Constructor for class `atp.AtpDaemonConsole`

AtpException() Constructor for class `atp.AtpException`
Constructs an `AtpException` with do detail message.

AtpException(String) Constructor for class `atp.AtpException`
Constructs an `AtpException` with the specified detail message.

AtpInputStream(InputStream) Constructor for class `atp.AtpInputStream`
Creates a new instance of `AtpInputStream`.

AtpNullOutputStream(PrintStream) Constructor for class `atp.AtpNullOutputStream`

AtpOutputStream(OutputStream) Constructor for class `atp.AtpOutputStream`
Creates a new instance of `AtpOutputStream`.

AtpRequestHandler() Constructor for class atp.AtpRequestHandler
Creates a new instance of this class.

AtpURLConnection(URL) Constructor for class atp.AtpURLConnection
Creates a new instance of this class.

AtpURLStreamHandler() Constructor for class atp.AtpURLStreamHandler

AtpURLStreamHandlerFactory() Constructor for class atp.AtpURLStreamHandlerFactory

atpUrlText() Method in class aglets.Log

B

BAD_GATEWAY Static variable in class atp.AtpOutputStream
Status code representing that the recipient, while acting as a gateway or proxy, received an invalid response from upstream server.

BAD_REQUEST Static variable in class atp.AtpOutputStream
Status code representing that the recipient was unable to understand the request message due to malformed syntax.

beep() Method in class aglets.util.MessageDialog
Rings a bell.

BETA Static variable in class aglets.Version
Beta version.

bounds() Method in class aglets.util.ImageComponent

C

callback(Slave, Object) Method in interface aglets.patterns.Master
Called when a slave has finished its job and wants to update its master with the result.

CANCEL Static variable in class aglets.util.MessageDialog

CENTER_ONLY_ONCE Static variable in class aglets.util.MessageDialog

checkMode(int) Method in class aglets.ThreadManager
Compares the requested execution mode to the current mode.

clearLog() Method in class aglets.patterns.Itinerary
Clears the log (remove all entries).

CLONE Static variable in class aglets.Log

clone() Method in class aglets.Aglet
Clones the aglet.

clone() Method in class aglets.AgletIdentifier
Creates an Aglet identifier with same family identity and unique dispatch identity.

cloneAglet(Aglet) Method in class aglets.AgletContext
Clones the specified aglet.

cloneThis() Method in class aglets.Aglet
Clones the aglet in the current execution context.

close() Method in class atp.AtpDaemonConsole
Closes this.

close() Method in class atp.AtpNullOutputStream
Closes the stream.

COMMON Static variable in class aglets.AgletOutputStream
Token indicating that a class written into this stream is common.

connect() Method in class atp.AtpURLConnection
Makes a comminucation link with the destination.

CONTENT_ENCODING Static variable in class atp.AtpOutputStream
A key value for CONTENT_ENCODING field in the message's header.

CONTENT_LENGTH Static variable in class atp.AtpOutputStream
A key value for CONTENT_LENGTH field in the message's header.

CONTENT_TYPE Static variable in class atp.AtpOutputStream
A key value for CONTENT_TYPE field in the message's header.

COOKIE Static variable in class atp.AtpOutputStream
A key value for COOKIE field in the message's field.

CREATE Static variable in class aglets.Log

create(AgletContext, URL, Object) Static method in class aglets.patterns.Messenger
Creates a messenger.

create(URL, String, AgletContext, Aglet, Itinerary, double, double, boolean, Object) Static method in class aglets.patterns.Notifier
Creates a notifier.

create(URL, String, AgletContext, Aglet, Itinerary, Object)
Static method in class aglets.patterns.Slave
Creates a slave.

create(URL, String, AgletContext, Aglet, Vector, Object)
Static method in class aglets.patterns.Slave
Creates a slave.

createAglet(URL, String, Object) Method in class aglets.AgletContext
Creates an instance of the specified aglet located at the specified URL.

createAglet(URL, String, Object, int) Method in class aglets.AgletContext
Creates an instance of the specified aglet located at the specified URL.

createURLStreamHandler(String) Method in class atp.AtpURLStreamHandlerFactory
Creates an instance of the AtpURLStreamHandler class.

CRLF Static variable in class atp.AtpDaemonConsole

CRLF Static variable in class atp.AtpOutputStream
A separator in the message's header.

D

DATE Static variable in class atp.AtpOutputStream
A key value for DATE field in the message's header.

DEF_CONF_DIR Static variable in class atp.AtpDaemonConsole

DEF_CONF_FILE Static variable in class atp.AtpDaemonConsole

DEFAULT Static variable in class aglets.AgletIdentifier
The default aglet identity.

dialog() Method in class aglets.AgletProxy
Request a dialog with the aglet.

disconnect() Method in class atp.AtpURLConnection
Disconnects a communication link.

DISPATCH Static variable in class atp.AtpInputStream
The method token for DISPATCH.

DISPATCH Static variable in class aglets.Log

dispatch(byte[]) Method in class `atp.AtpOutputStream`
Writes a dispatch request message with a body into this stream.

dispatchAglet(Aglet, URL) Method in class `aglets.AgletContext`
Dispatches the specified aglet to the specified destination.

dispatchHandler(URL) Method in class `aglets.patterns.Itinerary`
Dispatches an aglet to a specific URL.

dispatchThis(URL) Method in class `aglets.Aglet`
Dispatches this aglet to the location specified by the argument URL.

DISPOSE Static variable in class `aglets.Log`

disposeAglet(Aglet) Method in class `aglets.AgletContext`
Disposes the specified aglet.

disposeThis() Method in class `aglets.Aglet`
Removes this Aglet from the current execution context.

doCheck(Object) Method in class `aglets.patterns.Notifier`
This method should be overridden to specify the check method for this notifier.

doJob() Method in class `aglets.patterns.Slave`
This method should be overridden to specify the job of the slave.

E

enableAtp() Static method in class `atp.AtpURLStreamHandlerFactory`
Sets an instance of this class to the stream handler collection of the URL class.

equals(Object) Method in class `aglets.AgletCookie`
Compares two aglet cookies.

equals(Object) Method in class `aglets.AgletIdentifier`
Compares two aglet identifiers.

ERROR Static variable in class `aglets.Log`

error(String, String) Method in class `atp.AtpDaemonConsole`
Writes an error message string.

errorString(String, String) Method in class `atp.AtpDaemonConsole`
Creates an error message string.

errorWrite(String) Method in class `atp.AtpDaemonConsole`
Writes an error message into the output stream.

EXCEPTION Static variable in class `aglets.patterns.Notification`

exit() Method in class `atp.AtpDaemon`
Terminates this.

EXPIRY Static variable in class `aglets.patterns.Notification`

F

FETCH Static variable in class `atp.AtpInputStream`
The method token for `FETCH`.

fetch(String) Method in class `atp.AtpOutputStream`
Writes a fetch request message into this stream.

finalize() Method in class `aglets.AgletLoader`
Shouts when an `AgletLoader` object is caught by GC.

flush() Method in class `atp.AtpNullOutputStream`
Flushes the stream.

FORBIDDEN Static variable in class `atp.AtpOutputStream`
Status code representing that although the recipient understood the request message, it refused to fulfill it.

FREE Static variable in class `aglets.util.MessageDialog`

FROM Static variable in class `atp.AtpOutputStream`
A key value for `FROM` field in the message's header.

G

getAglet(AgletCookie) Method in class `aglets.AgletProxy`
Gets the aglet.

getAglet(AgletIdentifier) Method in class `aglets.AgletContext`
Gets the proxy for an aglet specified by its identity.

getAgletContext() Method in class `aglets.Aglet`
Gets the execution context that the aglet is currently running in.

getAgletGroup() Method in class `aglets.ThreadManager`
Gets the current aglet thread group.

getAgletIdentity() Method in class `aglets.ThreadManager`
Gets the aglet identity of the current aglet thread group.

getAgletProxies() Method in class `aglets.AgletContext`
Gets the aglet proxies in the current execution context.

getAgletURL() Method in class aglets.Log

getArrival() Method in class aglets.AgletProxy
Gets the time of proxy creation.

getBody() Method in class atp.AtpInputStream
Gets an input stream containing a body of the read data.

getBuild() Method in class aglets.Version
Gets the build number x.x.B

getButton(int) Method in class aglets.util.MessageDialog
Obtains dialog button

getByteCodeFromCache(String) Method in class aglets.AgletLoader
Gets a bytecode of a desired class from the class data cache.

getClassFromCache(String) Method in class aglets.AgletLoader
Gets a desired class from the class cache.

getClassName() Method in class aglets.AgletProxy
Gets the aglet's class name.

getCodeBase() Method in class aglets.AgletProxy
Gets the URL of the aglet's class.

getCurrentMode() Method in class aglets.ThreadManager
Gets the mode of the current aglet thread group.

getCurrentURL() Method in class aglets.patterns.Itinerary
Tells the URL of the current visited place on the tour.

getDate() Method in class aglets.Log

getDate() Method in class aglets.Version
Gets the date of the version.

getDefaultMode() Method in class aglets.ThreadManager
Returns the default execution mode.

getDispatchId() Method in class aglets.AgletIdentifier
Gets the dispatch identity.

getExpiration() Method in class aglets.Version
Gets the date of expiration.

getFamilyId() Method in class aglets.AgletIdentifier
Gets the family identity.

getHeaderField(String) Method in class atp.AtpInputStream
Gets a value of the header field read from this input stream with specified key.

getHeaderField(String) Method in class atp.AtpOutputStream
Gets a value of a header field specified by a key.

getHeaderField(String, String) Method in class atp.AtpInputStream
Gets a value of the header field read from this input stream with specified key.

getHeaderField(String, String) Method in class atp.AtpOutputStream
Gets a value of a header field specified by a key.

getHostingURL() Static method in class aglets.AgletContext
Returns the URL of the daemon serving all current execution contexts.

getHostName() Method in class atp.AtpDaemon
Gets the host domain name of the machine that is currently hosting the aglet listener.

getIdentity() Method in class aglets.Aglet
Returns the identity of this aglet.

getIdentity() Method in class aglets.AgletProxy
Gets the aglet's identity.

getIdentity() Method in class aglets.Log

getInputStream() Method in class atp.AtpURLConnection
Gets an input stream of the communication link.

getKind() Method in class aglets.patterns.Notification

getKind() Method in class aglets.Version
Gets the string that describes what is versioned.

getLine(FileInputStream) Method in class atp.AtpDaemonConsole
Gets a line from the configuration file.

getLongText() Method in class aglets.Version
Returns a long text representation of the version numbers: e.g., Alpha2b, Beta1, V2.0.

getMajor() Method in class aglets.Version
Gets the major version number M.x.x

getMajorText() Method in class aglets.Version
Returns a text representation of the major version number: e.g., Alpha, Beta, V2.

getMaster() Method in class aglets.patterns.Slave
Gets the URL of this slave's master.

getMessage() Method in class aglets.AgletProxy
Gets the current content of the Aglet's message line.

getMessage() Method in class aglets.Log

getMessage() Method in class aglets.patterns.Notification

getMessageType() Method in class atp.AtpInputStream
Gets message type read from this input stream.

getMethod() Method in class atp.AtpInputStream
Gets method token read from this input stream.

getMinor() Method in class aglets.Version
Gets the minor version number x.M.x

getNotifier() Method in class aglets.patterns.Notification

getNumber() Method in class aglets.patterns.Notification

getOutputStream() Method in class atp.AtpURLConnection
Gets an output stream of the communication link.

getPort() Method in class atp.AtpDaemon
Gets the port number that the aglet listener is currently listening to.

getProperty(String) Method in class aglets.Aglet
Gets the aglet property indicated by the key.

getProperty(String) Method in class aglets.AgletContext
Gets the context property indicated by the key.

getProperty(String) Method in class aglets.AgletProxy
Gets the aglet property indicated by the key.

getProperty(String, String) Method in class aglets.Aglet
Gets the aglet property indicated by the key and default value.

getProperty(String, String) Method in class aglets.AgletContext
Gets the context property indicated by the key and default value.

getProperty(String, String) Method in class aglets.AgletProxy
Gets the aglet property indicated by the key and default value.

getProperty(String, String) Method in class atp.AtpDaemon
Gets property.

getReasonPhrase() Method in class atp.AtpInputStream
Gets a reason phrase read from this input stream.

getReceiver() Method in class aglets.patterns.Notifier
Gets the URL of this notifier's receiver.

getShortText() Method in class aglets.Version
Returns a short text representation of the version numbers: e.g., A2b, B1, V2.0.

getStatusCode() Method in class atp.AtpInputStream
Gets a status code read from this input stream.

getTime() Method in class aglets.patterns.Notification

getType() Method in class aglets.Log

getURI() Method in class atp.AtpInputStream
Gets URI from this input stream.

getURL() Method in class aglets.Log

gotoByIndex(int) Method in class aglets.patterns.Itinerary
Dispatches an aglet to a new place with a specific index.

gotoErrorHandler() Method in class aglets.patterns.Itinerary
Dispatches an aglet to the handler.

gotoNext() Method in class aglets.patterns.Itinerary
Dispatches an aglet to the next available place in the tour.

gotoOrigin() Method in class aglets.patterns.Itinerary
Dispatches an aglet to the origin.

H

handleEvent(Event) Method in class aglets.util.MessageDialog
Handles the events.

handleRequest(AtpInputStream, AtpOutputStream) Method in class atp.handler.AgletsAtpRequestHandler

handleRequest(AtpInputStream, AtpOutputStream) Method in class atp.AtpRequestHandler
Handles a request message.

hashCode() Method in class aglets.AgletIdentifier
Returns an integer suitable for hash table indexing.

HELP Static variable in class aglets.util.MessageDialog

hide() Method in class aglets.util.MessageDialog

I

id Variable in class aglets.ObservableEvent

ImageComponent(Image) Constructor for class aglets.util.ImageComponent

ImageComponent(Image, Dimension) Constructor for class aglets.util.ImageComponent

imageUpdate(Image, int, int, int, int, int) Method in class aglets.util.ImageComponent

inError(Slave, Object) Method in interface aglets.patterns.Master
Called when a slave catches an exception.

initDefConfigFile(FileOutputStream) Method in class atp.AtpDaemonConsole
Writes initial strings in the configuration file.

initialize(AtpDaemon, String) Method in class atp.AtpDaemonConsole
Initializes an instance of this class.

initialize(Object) Method in class aglets.Aglet
Initializes the aglet.

initialize(Object) Method in class aglets.patterns.Messenger
Initializes the messenger.

initialize(Object) Method in class aglets.patterns.Notifier
Initializes the notifier.

initialize(Object) Method in class aglets.patterns.Slave
Initializes the slave.

initialize(URL) Method in class atp.handler.AgletsAtpRequestHandler

initialize(URL) Method in class atp.AtpRequestHandler

initializeCheck(Object) Method in class aglets.patterns.Notifier
This method should be overridden to specify the initial check performed by this notifier.

INSERT. Static variable in class aglets.ObservableEvent

inside(int, int) Method in class aglets.util.ImageComponent

instantiateAglet(URL, String) Static method in class `aglets.AgletLoader`
Instantiates the Aglet class specified by name.

instantiateObject(URL, String) Static method in class `aglets.AgletLoader`
Instantiates the object class specified by name.

INTERNAL_ERROR Static variable in class `atp.AtpOutputStream`
Status code representing that the recipient encountered an unexpected condition which prevented it from fullfiling the request.

isAtHandler() Method in class `aglets.patterns.Itinerary`
Tells whether the aglet is currently visiting the `Handler` host.

isAtLastDestination() Method in class `aglets.patterns.Itinerary`
Tells whether the aglet is currently visiting the last place in the tour.

isAtOrigin() Method in class `aglets.patterns.Itinerary`
Tells whether the aglet is currently located in the `Origin` host

isNull() Method in class `aglets.AgletIdentifier`
Tells whether this identifier is a null identifier.

Itinerary(URL) Constructor for class `aglets.patterns.Itinerary`
Creates an Itinerary with a single destination.

Itinerary(URL, Vector) Constructor for class `aglets.patterns.Itinerary`
Creates an Itinerary with a single destination.

Itinerary(Vector) Constructor for class `aglets.patterns.Itinerary`
Creates an Itinerary.

L

loadClass(String, boolean) Method in class `aglets.AgletLoader`
Loads a class specified by the param name.

Log(int, Aglet) Constructor for class `aglets.Log`

Log(int, Aglet, URL) Constructor for class `aglets.Log`

Log(int, String) Constructor for class `aglets.Log`

log(String, Date, Date,AtpInputStream) Method in class `atp.AtpDaemonConsole`
Writes a log message.

logString(String, Date, Date, AtpInputStream) Method in class atp.AtpDaemonConsole

Creates a log message string.

logToString() Method in class aglets.patterns.Itinerary

Converts the log to a string.

logWrite(String) Method in class atp.AtpDaemonConsole

Writes a log message into the output stream.

M

main(String[]) Static method in class atp.AtpDaemon

The main program of the ATP daemon.

makeThreadGroup(Aglet, int) Method in class aglets.ThreadManager

Creates the thread group that the aglet's thread is put into.

MESSAGE Static variable in class aglets.Log

MESSAGE Variable in class aglets.patterns.Notifier

The protected variable that carries any messages that should go along with the notification back to the subscriber.

MESSAGE Static variable in class aglets.ObservableEvent

message(Messenger, Object) Method in interface aglets.patterns.Receiver

Called when a messenger arrives with a message.

message(String) Method in class atp.AtpDaemonConsole

Writes a message into the output stream.

MessageDialog(Frame, Component, String, String, int, Object, boolean) Constructor for class aglets.util.MessageDialog

MessageDialog(Frame, String, String) Constructor for class aglets.util.MessageDialog

MessageDialog(Frame, String, String, Object) Constructor for class aglets.util.MessageDialog

MessagePanel(String, boolean) Constructor for class aglets.util.MessagePanel

MessagePanel(String, int, boolean) Constructor for class aglets.util.MessagePanel

`messageString(String)` Method in class `atp.AtpDaemonConsole`
Creates a message string.

`messageWrite(String)` Method in class `atp.AtpDaemonConsole`
Writes a message into the output stream.

`Messenger()` Constructor for class `aglets.patterns.Messenger`

`move(int, int)` Method in class `aglets.util.ImageComponent`

`MOVED` Static variable in class `atp.AtpOutputStream`
Status code representing that the requested resource is no longer at the recipient.

N

`NO_COOKIE` Static variable in class `aglets.AgletCookie`
The static `NoKey` method generates an “authentication-less” key.

`NONE` Static variable in class `aglets.ObservableEvent`

`NOT_FOUND` Static variable in class `atp.AtpOutputStream`
Status code representing that the recipient could not find the requested resource.

`NOT_IMPLEMENTED` Static variable in class `atp.AtpOutputStream`
Status code representing that the recipient does not support the functionality required to fulfill the request.

`NOTIFICATION` Static variable in class aglets.patterns.Notification

`Notification(int, URL, Date, int, Object)` Constructor for class `aglets.patterns.Notification`
This constructor is used to construct the `Notification` to be sent.

`Notifier()` Constructor for class `aglets.patterns.Notifier`

`NULL` Static variable in class `aglets.AgletIdentifier`
The null aglet identity.

O

`ObservableEvent(int)` Constructor for class `aglets.ObservableEvent`

`ObservableEvent(Object, int)` Constructor for class `aglets.ObservableEvent`

`OKAY` Static variable in class `atp.AtpOutputStream`
Status code representing that the request has succeeded.

OKAY Static variable in class `aglets.util.MessageDialog`

onArrival() Method in class `aglets.Aglet`
Called when the aglet is arriving from a remote server.

onDialog() Method in class `aglets.Aglet`
Used to request the aglet to enter into a dialog with the user.

onDispatching(URL) Method in class `aglets.Aglet`
Called when the Aglet is dispatched.

onDisposal() Method in class `aglets.Aglet`
Called when the aglet is disposed.

onRetraction(URL) Method in class `aglets.Aglet`
Called when the aglet is retracted from a remote server.

openConnection(URL) Method in class `atp.AtpURLStreamHandler`
Opens a connection to the object referenced by the URL argument.

P

pack() Method in class `aglets.util.MessageDialog`

paint(Graphics) Method in class `aglets.util.ImageComponent`

paint(Graphics) Method in class `aglets.util.MessagePanel`

popup(Frame) Method in class `aglets.util.MessageDialog`
Pops up the dialog window so that it will be located at the center of the frame.

popup(int) Method in class `aglets.util.MessageDialog`
Pops up the dialog window according to the location given as a parameter.

preferredSize() Method in class `aglets.util.ImageComponent`

printStringCenter(String, Graphics) Method in class `aglets.util.ImageComponent`

propertyKeys() Method in class `aglets.Aglet`
Enumerates all the property keys.

propertyKeys() Method in class `aglets.AgletContext`
Enumerates all the property keys.

propertyKeys() Method in class `aglets.AgletProxy`
Enumerates all the property keys.

putByteCodeIntoCache(String, byte[]) Method in class `aglets.AgletLoader`
Puts a bytecode of a class into the class data cache.

putClassIntoCache(String, Class) Method in class
aglets.AgletLoader
Puts a class into the class cache.

R

read() Method in class atp.AtpInputStream
Reads a next byte of data from this input stream.

RECEIVE Static variable in class aglets.Log

receiveAglet(Aglet) Method in class aglets.AgletContext
Receives an aglet.

receiveAglet(Aglet, int) Method in class aglets.AgletContext
Receives an aglet.

RECIPIENT_APPLICATION Static variable in class atp.AtpOutputStream
A key value for RECIPIENT_APPLICATION field in the message's header.

REMOVE Static variable in class aglets.ObservableEvent

removeByteCodeFromCache(String) Method in class
aglets.AgletLoader
Removes a bytecode from the class data cache.

reset() Method in class aglets.patterns.Itinerary
Resets an aglet's tour as if the tour have not yet been started.

resolveClass(String) Method in class aglets.AgletInputStream
Resolves a class specified by classname.

resolveConfigFile(String) Method in class atp.AtpDaemonConsole
Reads the config file and get paramters.

respond(int) Method in class atp.AtpOutputStream
Writes a respond message into this stream.

respond(int, byte[]) Method in class atp.AtpOutputStream
Writes a respond message into this stream.

restart() Method in class atp.AtpDaemon
Called to restart the aglet listener on the same port.

restart(int) Method in class atp.AtpDaemon
Called to restart the aglet listener on a new port.

RESULT Variable in class aglets.patterns.Slave
The protected variable that keeps the result of task that this slave performs.

RETRACT Static variable in class atp.AtpInputStream
The method token for RETRACT.

RETRACT Static variable in class aglets.Log

retract(String) Method in class atp.AtpOutputStream
Writes a retract request message into this stream.

retract(String, String) Method in class atp.AtpOutputStream
Writes a retract request message into this stream.

retractAglet(URL) Method in class aglets.AgletContext
Retracts the Aglet specified by its url: atp://host-domain-name/[user-name]#aglet-identity.

retractAglet(URL, int) Method in class aglets.AgletContext
Retracts the Aglet specified by its URL.

REVERT Static variable in class aglets.Log

revertAglet(URL, AgletIdentifier) Method in class aglets.AgletContext

RP_BAD_GATEWAY Static variable in class atp.AtpOutputStream
Reason phrase representing that the recipient, while acting as a gateway or proxy, received an invalid response from upstream server.

RP_BAD_REQUEST Static variable in class atp.AtpOutputStream
Reason phrase representing that the recipient was unable to understand the request message due to malformed syntax.

RP_FORBIDDEN Static variable in class atp.AtpOutputStream
Reason phrase representing that although the recipient understood the request message, it refused to fulfill it.

RP_INTERNAL_ERROR Static variable in class atp.AtpOutputStream
Reason phrase representing that the recipient encountered an unexpected condition which prevented it from fullfiling the request.

RP_MOVED Static variable in class atp.AtpOutputStream
Reason phrase representing that the requested resource is no longer at the recipient.

RP_NOT_FOUND Static variable in class atp.AtpOutputStream
Reason phrase representing that the recipient could not find the requested resource.

RP_NOT_IMPLEMENTED Static variable in class `atp.AtpOutputStream`
Reason phrase representing that the recipient does not support the functionality required to fulfill the request.

RP_OKAY Static variable in class `atp.AtpOutputStream`
Reason phrase representing that the request has succeeded.

RP_SERVICE_UNAVAILABLE Static variable in class `atp.AtpOutputStream`
Reason phrase representing that the recipient is currently unable to handle the request due to a temporary overloading of the recipient.

run() Method in class `aglets.Aglet`
This method is the entry point for the aglet's own thread.

run() Method in class `atp.AtpDaemon`
The entry point for the daemon thread.

run() Method in class `atp.AtpDaemonConsole`
This method is called from the console thread.

run() Method in class `aglets.patterns.Messenger`
Universal entry point for the messenger's execution thread.

run() Method in class `aglets.patterns.Notifier`
Universal entry point for the notifier's execution thread.

run() Method in class `aglets.patterns.Slave`
Universal entry point for the slave's execution thread.

S

SENDER_APPLICATION Static variable in class `atp.AtpOutputStream`
A key value for `SENDER_APPLICATION` field in the message's header.

SERVICE_UNAVAILABLE Static variable in class `atp.AtpOutputStream`
Status code representing that the recipient is currently unable to handle the request due to a temporary overloading of the recipient.

setButtons(int) Method in class `aglets.util.MessageDialog`
Specifies which buttons should appear on the bottom of window.

setCallbackComponent(Component) Method in class `aglets.util.MessageDialog`

setDestruction(boolean) Method in class `aglets.patterns.Itinerary`
Defines whether to dispose an aglet after it is dispatched or not.

setHandler(URL) Method in class aglets.patterns.Itinerary
Defines the URL of an host to which the aglet can be dispatched in cases of unexpected events during its tour.

setHeaderField(String, String) Method in class atp.AtpOutputStream
Sets a value of a header field specified by a key.

setHostingURL(URL) Static method in class aglets.AgletContext
Sets the URL of the daemon serving all current execution contexts.

setMessage(String) Method in class aglets.Aglet
Sets the message line of this Aglet.

setMessage(String) Method in class aglets.util.ImageComponent

setMessage(String) Method in class aglets.util.MessageDialog
Sets the message.

setNumAllowedFailures(int) Method in class aglets.patterns.Itinerary
Defines maximum number of allowed failures to visit a place during a tour.

setNumRetries(int) Method in class aglets.patterns.Itinerary
Defines maximum number of retries to dispatch an aglet to a new place.

setProperty(String) Method in class atp.AtpDaemonConsole
Sets properties to the atp daemon.

setProperty(String, String) Method in class atp.AtpDaemon
Sets property of this.

setTraveller(Aglet) Method in class aglets.patterns.Itinerary
Binds an aglet (for the tour).

size() Method in class aglets.util.ImageComponent

Slave() Constructor for class aglets.patterns.Slave

start() Method in class atp.AtpDaemon
Called to start the aglet listener.

start() Method in class atp.AtpDaemonConsole
Starts this.

stop() Method in class atp.AtpDaemon
Called to stop the aglet listener.

T

target Variable in class aglets.ObservableEvent

ThreadManager() Constructor for class aglets.ThreadManager
Constructs the thread manager.

toMessageText() Method in class aglets.Log

toString() Method in class aglets.AgletIdentifier
Returns a human readable form of the aglet identifier.

toString() Method in class aglets.Log

toString() Method in class aglets.Version
Returns a human readable form of the version object.

U

UNDEFINED Static variable in class aglets.Log

UPDATE Static variable in class aglets.ObservableEvent

urlText() Method in class aglets.Log

Utils() Constructor for class aglets.patterns.Utils

V

VERSION Static variable in class aglets.Aglet
Subclasses can set their version number.

Version(String, int, int, int, Date) Constructor for class aglets.Version
Creates a version object which will never expire.

W

waitForDisposal() Method in class aglets.util.MessageDialog
Waits until the dialog window is disposed.

write(byte[]) Method in class atp.AtpNullOutputStream

write(byte[], int, int) Method in class atp.AtpNullOutputStream

write(int) Method in class atp.AtpNullOutputStream

write(int) Method in class atp.AtpOutputStream
Writes a byte into this stream.

glossary of terms

Cross references to other terms in the glossary are printed in italics when occurring in descriptions. Some terms have more than one entry based on differing usage in the different agent systems. Any term that is specific to a given language is noted as such at the end of the definition.

Access Right

Operations on agents, such as terminating them, require the executing agent to possess the access right over the concerned agent. An agent possesses the access right over all agents of its own *group* and of any child *groups* thereof. (Ara)

Agent

A mobile agent is a program with the ability to move during execution, while preserving its identity and state. For the sake of uniformity, even programs which do not really move are subsumed under this term. Agents bear a globally unique immutable name. Each agent is the (possibly single) member of one *group* at a time, which bears an *allowance* limiting its resource accesses. Agents are executed as parallel *processes*, usually by an *interpreter*.

Agent (2)

Agents bear a globally unique immutable name. Each agent is the (possibly single) member of one *group* at a time, which bears an *allowance* limiting its resource accesses. Agents are executed as parallel *processes*, usually by an *interpreter*. (Ara)

Allowance

An allowance is a vector of permissions for various system resources, such as files, CPU time, or disk space. The elements of such a vector constitute quantitative permissions (e.g. for CPU time) or qualitative ones (e.g. for network domains to where connection is allowed). Each agent is equipped with a global allowance for its life time and may be further restricted by a local allowance while staying at a certain *place*. An agent shares both its global and local allowance with the other member agents of its *group*. (Ara)

Attribute An attribute contains an object's externally visible characteristics. An object can get or set its own attributes and can get the public attributes of any other object. (Telescript)

API An Application Programming Interface is the defined command that can be used to access the system resources that the programmer is working with.

Authority The authority of an agent or place is the individual or organization in the real world that it represents. Agents should neither withhold nor falsify their authorities.

Checkpoint A record of the complete *internal state* of an agent at some time. The concerned agent may be restored from the checkpoint, to resume its execution from this state. (Ara)

Class Classes, or aspects of classes, are part of all of the languages in the book, but each one is slightly different. In general, a class is a set of software instructions that contains both an interface and a related implementation. Classes do not actually perform operations but are used as templates for the agents actually doing the work.

Client An agent which has *met* a *service point*. The client may submit a *request* to the *service point* in order to receive a *reply* to the *request*. An agent may play the role of client and *server* at many *service points* at a time. (Ara)

Compiled Agent An agent which is compiled to native machine code and executed directly, as opposed to being *interpreted*. Compiled agents must be absolutely trustworthy, since they are able to subvert the security measures of the system. Compiled agents cannot usually *migrate*, except in certain cases where their source code is available. (Ara)

Core The central part of an Ara system, implementing the basic concepts such as *agents*, *allowances*, *service points*, *migration* etc. Any access from an application agent to the host system or to another agent is mediated by the core for reasons of security and portability. The core treats agents independently of their programming language, using assistance from the language *interpreters* for language-specific tasks. (Ara)

Exception When a value cannot be returned to the caller operation, an exception can be thrown. The exception can be caught by the caller or caught at the higher level of the agent's or server's systems. An error can be considered a safe error.

External State The relations of an agent to external objects, such as other agents, *service points*, or files. These relations cannot be preserved across *migration* or *checkpointing*, as such external objects cannot be warranted to exist where the agent completes its *migration* or when it is restored from a *checkpoint*. (Ara)

Fetch A *server* may fetch *requests* submitted to a *service point*. Fetched requests are tagged with the name of the requesting *client*. The fetch operation optionally waits until there are any *requests* submitted. (Ara)

GUI Graphical User Interface. A term given to the windowing systems pioneered on the Xerox PARC and Macintosh computer systems.

Group A group comprises a set of agents with a common *allowance*. Newly created agents join the group of their creator by default, but they may also be created to form a new, separate group, called a child of the creator's group. A child group receives a share of its parent's *allowance* and returns (what is left of) this when it becomes empty again. (Ara)

ID An ID is a globally unique and immutable machine-generated identification of an Ara core object. Currently, *agents* and *service points* bear IDs. Agents use IDs to name objects as arguments to operations on them. (Ara)

Implementation An object's implementation is made up of the object's *properties* and *methods*. (Telescript)

Interaction *Agents* usually interact locally, by *meeting* at a *service point*. There will also be a facility to pass messages between remote agents in a future version of Ara. (Ara)

Interface The *operations* and *attributes* associated with an *object.* (Telescript)

Internal State The complete execution state of an agent, except its *external state.* This comprises the agent's data, code, and execution context. (Ara)

Interpreter A mobile agent is executed within an interpreter for its programming language, for reasons of security and portability. An Ara system may contain interpreters for several languages on top of the common *core.* Besides executing the code of the interpreted agent, an interpreter must be adapted to the *core* in order to support it in the portable and secure execution of the agent. Currently, interpreters for the Tcl and C/C++ languages have been adapted to the Ara *core*, and an adaptation of Java is on the way. (Ara)

Java Java is a transportable language developed by Sun Microsystems. (Aglets)

JDK The Java Development Kit is the set of classes that are the core language for the Java language. It is free and can be obtained from Sun Microsystems. Information on where and how to obtain it are on the CD-ROM. (Aglets)

Meeting When agents are in the same place they can meet. When agents meet they can call one another's procedures, allowing them to exchange information or request operations.

Meeting (2) An agent may meet a *service point* at the local *place* under its symbolic name to become a *client* at this *service point.* The meet operation optionally waits until such a *service point* exists. A *client* may leave the service point again, resolving its relation to the *service point.* (Ara)

Method A procedure that performs an operation that can get or set an attribute. Methods can have variables that constitute its dynamic state. (Telescript)

Migration An act of active motion of an executing agent from one *place* to another, while preserving the complete *internal state* of the agent. In particular, its execution context is preserved, that is, the agent resumes directly after the migration statement in its program. The agent's *external state* cannot be preserved. A migrating agent takes along some *allowance* and leaves its group behind, forming a new group at the destination. It is possible to make another agent migrate; however, this agent should be prepared to make sense of this. (Ara)

Mobility The ability for *agents* to move from site to site, or server to server, in order to accomplish defined goals. A key feature of mobile agents is the ability to continue their operations on the new machine after the last instruction performed on the previous machine; the agents have state and carry it with them.

Mobility (2) *Agents* are the only mobile objects in Ara. Mobility therefore appears in the form of agent *migration*. (Ara)

Object Any piece of information in Telescript. All objects contain both an external interface and an internal implementation. (Telescript)

Operation Any task that an object can perform. An object can request its own operations and the public operations of other objects. Operations can accept other objects as arguments and can return a single object as the result. Operations can throw exceptions instead of returning an object. (Telescript)

Permit Permits grant agents the right to execute certain instructions or to use certain resources on a machine.

Place A place is a virtual location within an agent system that may be based on a physical system but can contain any number of services for agents to request.

Place (2) An agent is always either staying at some place or *migrating* between two of them. A place establishes a domain of logically related services under a common security policy governing all agents at that place, by deciding on the admission of an agent attempting to *migrate* to this place, possibly imposing a local *allowance* on that agent for the time of its stay. Places bear globally unique names. The current Ara implementation provides only one place per system, admitting any migrating agent without restrictions. (Ara)

Platform A general term that refers to the machine, operating system, and application level software that the software under consideration is running on. As an example, MacOS, Windows, and UNIX are various operating system platforms.

Process Agents running on an Ara system are executed as quasi-parallel processes, scheduled using a time-slicing scheme which is preemptive at the level of program instructions, but nonpreemptive at the level of processor instructions. The current scheduling policy is round robin without priorities. Processes are implemented using a fast thread package in the *core.* (Ara)

Property Owned by an object's *implementation*, the property is one of an object's internal characteristics. As with all things in Telescript, a property is itself an object. (Telescript)

Reject A *server* may *reject* a *request* submitted by a *client* instead of *replying* to it. Rejection will cause the *client's request* submission to fail. (Ara)

Reply A message of arbitrary format returned as the answer to a *request* by a *server.* The *server* may also *reject* the *request* instead of replying. (Ara)

Request A message of arbitrary format submitted to a *service point* by a *client.* The submission operation will either return the *reply* to the request or a *rejection.* (Ara)

Root Agent The initial agent running when an Ara system starts up. In the current implementation, the root agent executes Tcl code read from the input stream (a terminal in interactive mode), and termination of the root agent terminates the system. (Ara)

RPC Remote Procedure Calling is a dominant method used for computer-computer communications that allows for one computer to call procedures on another computer. Each message sent between machines on the network either requests or acknowledges a remote procedure's operation.

RP Remote Programming, a newer alternative to RPC, allows for computer-computer communications by transporting all data and operations from the calling computer to the remote computer. At the remote computer, all operations are performed without communication with the calling computer. All results are returned to the calling computer once operations are complete.

Server An agent which has created a *service point*. The server may *fetch requests* submitted to the *service point* and *reply* to or *reject* them. The server may delete the *service point* at any time, which implies closing it and *rejecting* all *requests* which had been submitted, but not yet *replied* to. An agent may play the role of server and *client* at many *service points* at a time. (Ara)

Service Point A service point is a meeting point for agents under a unique symbolic name, providing synchronous agent interaction. A service point has one *server* agent and arbitrarily many *client* agents. A service point may be temporarily closed and reopened by the *server*, affecting its behavior concerning attempts to *meet* or submit *requests*: Closed service points appear nonexistent to *meet* attempts, and *request* submission attempts are *rejected. Requests* already submitted, but not yet *replied* to are not affected by closing. A service point is always tied to a *place.* (Ara)

System Process Various tasks within the Ara system are performed by *processes* with special privileges, for example, to access the host operating system without mediation through the *core*. Such system processes are usually *compiled.* Apart from this, they are treated as any other *process* by the *core*. (Ara)

Tcl Tool Command Language was developed by Sun Microsystems and is intended as a high-level scripting language. (Agent Tcl)

Time Agents rely on the clock of the local system, on an internal clock, or on absolute time when executing. Each of these types of time are critical to varying aspects of agents and each possesses inherent abilities or inaccuracies.

Time (2) Each Ara system has a local clock, and agents may inquire the time on that clock, or wait for a certain time. Currently, the clock is implemented purely logically, that is, without access to a physical clock, which makes it somewhat inaccurate. (Ara)

bibliography

Arnold, K., and J. Gosling. *The Java Programming Language*. Reading, Mass: Addison-Wesley, 1996.

Apple Computer, Inc. "Human Interface Guidelines." Reading, Mass.: Addison-Wesley, 1994.

Avrahami, G., K.P Brooks, and M.H. Brown. "A Two-View Approach to Constructing User Interfaces," *Computer Graphics* (1989): 23 (3):137–146.

Badrinath, B.R., A. Bakre, T. Imielinski, and R. Marantz. "Handling Mobile Clients: A Case for Indirect Interaction," *Proceedings of the 4th Workshop on Workstation Operating Systems*. (IEEE Computer Society Press, 1993): 91–97.

Bal, H.E., J.G. Steiner, and A.S. Tanenbaum. "Programming Languages for Distributed Computing Systems," *ACM Computing Surveys* (1989): 21(3).

Bal, H., et al. "Orca: A Language for Distributed Processing," *SIGPLAN Notices* (May): 25 (5): 17–24.

Berners-Lee, T., et al. "The World Wide Web," *Communications of the ACM* (August 1994): 37(8): 76–82.

Berners-Lee, T., R. Fielding, and H. Frystyk. "Hypertext Transfer Protocol—HTTP/1.0," (1996). *Internet RFC 1945*. Available from `http://ds.intemic.net/rfc/rfc1945.txt`

Berners-Lee, T., M. Masinter, and M. McCahill. "Uniform Resource Locators (URL)," 1994. Internet RFC 1738 available from `http://ds.internic.net/rfc/rfc1738.txt`

Bharat, K.A. and L. Cardelli. "Migratory Applications," *Systems Research Center Report*, Digital Equipment Corporation (February 1996).

Bharat, K., and M.H. Brown. "Building Distributed Multi-User Applications By Direct Manipulation," *Proceedings of the ACM Symposium on User Interfaces Software and Technology* (Marina Del Rey, 1994): 71–82.

Bharat, K., and P. Sukaviriya. "Animating User Interfaces with Animation Servers," in *Proceedings of the UIST' 93.*: 69–79.

Birrell, A.D., G. Nelson,, S. Owicki, and E. Wobber, "Network Objects,"*Proceedings of the 14th Symposium on Operating Systems Principles* (1993).

Birrell, A. D., and B.J. Nelson. "Implementing Remote Procedure Calls," *ACM Transactions on Computer Systems* (1984): 2(1): 39–59.

Black, A., et al. "Distribution and Abstract Types in Emerald," in *IEEE Transactions on Sofiware Engineering* (January 1987): SE-13(1):65–76.

Black, A., N. Hutchinson, E. Jul, and H. Levy. "Fine-Grained Mobility in the Emerald System." *ACM Transactions on Computer Systems* (1988) 6(1): 109–133.

Borenstein,N.S. and M. Rose. "MIME Extensions for Mail-Enabled Applications: Application/Safe-Tcl and Multipart/enabled-mail," draft, Bellcore, Dover Beach Consulting, September 1993.

Borenstein, N.S. and M. Rose. "Safe Tcl." Available at `ftp://ftp.fv.com/pub/code/other/safe-tcl.tar.Z`

Cai, T., P.A. Gloor, and S. Nog. "DartFlow: A Workflow Management System on the Web Using Transportable Agents," *PCS-TR96-283* (May 1996): Department of Computer Science, Dartmouth College.

Cardelli, L. "A Language with Distributed Scope," *Computing Systems* (1995): 8(1), 27–59.

Chess, D., B. Grosof, and C. Harrison. "Itinerant Agents for Mobile Computing," *Research Report RC-20010*, IBM Th. J. Watson Research Center (1995).

Chevalier, P.Y. and B. Thomsen. "Mobile Service Agents," 1995. Available from `http://www.ecrc.de/research/dc/msal`

Coen, M.D., "SodaBot: A Software Agent Environment and Construction System," *Proceedings of the CIKM Workshop on Intelligent Information Agents, Third International Conference on Information and Enowledge Management* (CIKM 94).

Condict, M., D. Milojicic, F. Reynolds and D. Bolinger. "Towards a World-Wide Civilization of Objects," to appear in *Proceedings of the 7th ACM SIGOPS European Workshop*, 1996 Connemara, Ireland.

Crocker, D.H. "Standard for the Format of ARPA Internet Text Messages," Internet RFC 822, 1982. Available from `http://ds.internic.net/rfc/rfc822.txt`

Cypher, A. "EAGER: Programming Repetitive Tasks by Example," *Proceedings of CHI '91*: 33– 39.

Cypher, A., ed., *Watch What I Do—Programming bv Demonstration,* MIT Press, 1993.

Emtage, A. and P. Deutsch. "Archie: An Electronic Directory Service for the Internet," *Proceedings of the U.S. ENIX Winter 1992 Conference*: 93–110.

Falcone, J.R. "A Programmable Interface Language for Heterogeneous Distributed Systems," *ACM Transactions on Computer Systems* (1987): 5(4).

Forman, G. and J. Zahorjan, J. "The Challenges of Mobile Computing," *Technical Report CSE-93-11-03,* University of Washington. 1994.

Foner, L. "What's an Agent, Anyway? A Sociological Case Study," *MIT Media Lab Agents Memo 93-01*, Massachusetts Institute of Technology, Cambridge Mass., 1993. Available at `http://foner.www.media.mit.edu/people/foner/Julia/`

Franklin, S. and A. Graesser. "1s it an Agent, or Just a Program? A Taxonomy for Autonomous Agents," Institute for Intelligent Systems, University of Memphis, 1996. Available from `http://www.msci.memphis.edu/~franklin/AgentProg.html`

Gaines, R.S., "Dixie Language Design and Interpreter Issues," *Proceedings of the USENIX Symposium on Very High Level Languages (VHLL),* 1994.

General Magic, Inc. 'The Telescript Language Reference," Sunnyvale, Calif., 1995.

Gifford, D. K., and J.W. Stamos. "Remote evaluation," *ACM Transactions on Programming Languages and Systems* (1990): 12(4): 537–565.

Goldszmidt, G., and Y.Yemini. "Distributed Management by Delegating Mobile Agents," *Proceedings of the 15th International Conference on Distributed Computing Systems.* Vancouver, Canada: 1995. Available from: `http://www.cs.columbia.edu/~german/papers/icdcs95.ps.Z`

Goldberg, D., D. Nichols, B. Oki, and D. Terry. "Using Collaborative Filtering to Weave an Information Tapestry," *Communications of the ACM*, (1992): 35(12): 61–70.

Gosling, J., and H. McGilton. "The Java Language Environment: A White Paper." Sun Microsystems, (1995).

Gray, R., D. Kotz, S. Nog, D. Rus, and G. Cybenko. "Mobile Agents for Mobile Computing," *PCS-TR96-285*, Department of Computer Science, Dartmouth College (1996).

Gray, R.S. "Agent Tcl: A Transportable Agent System," *Proceedings of the CIKM Workshop on Intelligent Information Agents* (1995).

Gray, R.S. "Agent Tcl: A Flexible and Secure Mobile-Agent System," *Proceedings of the Fourth Annual Tcl/Tk Workshop* (1996).

Harker, K.E. "TIAS: A Transportable Intelligent Agent System," *PCS-TR95-258*, Department of Computer Science, Dartmouth College, 1995.

Harrison, C., D. Chess, and A. Kershenbaum. "Mobile Agents—Are They a Good Idea?", *RC-19887*, IBM Th. J. Watson Research Center, 1994.

Horton, M,R. and R. Adams. "Standard for Interchange of USENET Messages," *Internet RFC 1036*, AT&T Bell Laboratories and Center for Seismic Studies, 1987. Available from `http://ds.internic.net/rfc/rfc1036.txt`

Hohl, F. "Konzeption eines einfachen Agentensystems und Implementation eines Prototyps," diploma thesis no. 1267, Department of Computer Science, University of Stuttgart, Germany, 1995.

Hughes, M., C. Hughes, M. Shoffner, and M. Winslow, M. *Java Network Programming* Manning Publications: 1997.

Johansen, D., R. van Renesse, and F.B. Scheidner. "Operating System Support for Mobile Agents,"*Proceedings of the 5th IEEE Workshop on Hot Topics in Operating Systems.* (1995), 42–45.

Johansen, D., R. van Renesse, and F.B. Schneider. "An Introduction to the TACOMA Distributed System," *Technical Report 95-23*, Dept. of Computer Science, University of Tromsø Norway, 1995. Available from `http://www.cs.uit.no/Lokalt/Rapporter/Reports/9523.html`

Kantor, B. and P. Lapsley. "Network News Transfer Protocol," *Internet RFC 977*, University of California, San Diego and University of California, Berkeley, February 1986. Available from `http://ds.internic.net/rfc/rfc977.txt`

Kaufman, C., R. Perlman, and M. Speciner. *Network Security: Private Communication in a Public World.* New Jersey: Prentice-Hall, 1995.

Kotay, Keith, and David Kotz, "Transportable Agents," *Proceedings of the CIKM Workshop on Intelligent Information Agents, Third International Conference on Information and Knowledge Management*, 1994.

Li, W. "Java-To-Go: Itinerative Computing Using Java," 1996. Available from `http://ptolemy.eecs.berkeley.edu/dgm/javatools/java-to-go/`

Lingnau, A., O. Drobnik, and P. Dömel. "An HTTP-based Infrastructure for Mobile Agents," in *Proceedings of the 4th International WWW Conference*, Boston, Mass. 1995. Available from http://www.w3.org/pub/Conferences/WWW4/Papers/150/

Lucco, S., O. Sharp, and R. Wahbe. "Omniware: A Universal Substrate for Web programming," *Proceedings of the 4th International WWW Conference*, Boston, Mass. 1995.

Manasse, M.S., and G. Nelson, "Trestle Reference Manual," *Research Report 68*, Digital Equipment Corporation, Systems Research Center, 1991.

Mayfield, J., Y. Labrou, and T. Finin. "Desiderata for Agent Communication Languages," *Proceedings of the AAAI Symposium on Information Gathering from Heterogeneous, Distributed Environments*, 1995 Spring Symposium. Available from http://www.cs.umbc.edu/kqml/papers/desiderata-acl/root.html

Niemeyer, P., and J. Peck. *Exploring Java*. O'Reilly R Associates, 1996.

Nog, S., S. Chawla, and D. Kotz, "An RPC Mechanism for Transportable Agents," *PCS-TR96-280*, Department of Computer Science, Dartmouth College, 1996.

Object Management Group. "CORBA 2.0 Specification," *OMG Document PTC96-03-04*, I996. C-]. Available from http://www.omg.org/docs/ptc/96-03-04.ps

Ousterhout, J. K. "Scripts and Agents: The New Software High Ground," keynote address at the 1995 USENIX winter conference.

Ousterhout, J.K., *Tcl and the Tk Toolkit*. Reading, Mass.: Addison-Wesley, 1994.

Peine, H. "The ARA Project," distributed Systems Group, Department of Computer Science, University of Kaiserlautern, 1996.

Perret, S. and A. Duda. "Mobile Assistant Programming for Efficient Information Access on the WWW," *Proceedings of the 5th WWW Conference*, Paris, France, 1996. Available from http://www5conf.inria.fr/fich_html/papers/P42/Overview.html

Powell, M. and B. Miller. "Process Migration in DEMOS/MP," *Proceedings of the 9th ACM Symposium on Operating System Principles*, (1983): 110–119.

Price, R., ed., "Proposed ISO/IEC International Standard for HTML" *Proposal of the JTC1 Joint Technical Committee Shared by the International Organization for Standardization (ISO) and the International Electrotechnical Commission (IEC)*, 1996. Available from http://www.w3.org/pub/WWW/MarkUp/JTC1-SC29/Overview.html

Rosenschein, J. S. and G. Zlotkin. *Rules of Encounter*. MIT Press, 1994.

Rouaix, F. "A Web Navigator with Applets in Caml," *Proceedings of the 5th WWW Conference*, 1996 Available from `http://www5conf.inria.fr/fich_html/papers/P41/Overview.html`

Sah, A. "TC: An Efficient Implementation of the Tcl Language," *Technical Report UCB-CSD-94- 812*. University of California at Berkeley, May 1994.

Sah, A., J. Blow, and B. Dennis. "An Introduction to the Rush Language," *Proceedings of the 1994 Tcl Workshop*.

Scherer, M. *Ein Laufzeitsystem fur mobile Agenten*, diploma thesis, Department of Computer Science, University of Kaiserslautern, Germany, 1995.

Singhal, M. and N.G. Shivaratri. *Advanced Concepts in Operating Systems: Distributed, Database and Multiprocessor Operating Systems*. New York: McGraw-Hi1l, 1994.

Sheth, B., and P. Maes. "Evolving Agents for Personalized Information Filtering," *Proceedings of IEEE Conference on AI for Applications*, 1993.

Shneiderman, B. "Direct Manipulation: A Step Beyond Programming Languages," *Computer*, 1983: 16(8), 57–68

Stamos, J.W. and W.K. Gifford. "Remote Evaluation," *ACM Transactions on Programming Languages and Systems*: October 1990: 12(4): 537–565.

Stolpmann, T. "MACE—Mobile Agent Code Environment (Mace—Eine abstrakte Maschine als Basis mobiler Anwendungen)," diploma thesis, Department of Computer Science, University of Kaiserslautern, Germany, 1996; German text and English summary available from `http://www.uni-kl.de/AG-Nehmer/Ara/mace.html`

Stoyenko, A.D. "SUPRA-RPC: Subprogram Parameters in Remote Procedure Calls," *Software—Practice and Experience*, (January 1994): 24(1):27–49.

Stroustrup, B. *The C++ Programming Language*, 2nd ed. Reading, Mass.: Addison-Wesley, 1990.

Sun Microsystems, *Hot Java Browser: A White Paper*. Sun Microsystems, 1994.

Sun Microsystems, *The Java Language: A White Paper*. Sun Microsystems, 1994.

Tardo, J. and L. Valente. "Mobile Agent Security and Telescript," *Proceedings of the 41st International Conference of the IEEE Computer Society CompCon '96*, February 1996.

Thistlewaite, P. and S. Ball. "Active FORMs," *Proceedings of the 5th WWW Conference,* May 1996 Paris, France. Available from `http://www5conf.inria.fr/fich_html/papers/P40/Overview.html`

Thomsen, B., L. Leth, F. Knabe and P. Chevalier. *Mobile Agents,* European Computer-Industry Research Centre, 1995.

Tschudin, C. "Messengers and Object-Oriented Agents," Position paper for the workshop "Objects and Agents" at the *8th European Conference on Object-Oriented Programming,* Aarhus, Denmark, 1995; available from `http://cuiwww.unige.ch/OSG/Vitek/Agents/christian.ps.gz`

Vittal, J. *Active Message processing: Messages as Messengers.* Computer Message Systems, North-Holland Publishing Company, 1981.

Wahbe, R., S. Lucco, T. Anderson, S.I. Graham. "Efficient Software-Based Fault Isolation," *Proceedings of the 14th ACM Symposium on Operating Systems Principles,* Asheville, N.C., 1993.

Welch, B.B., *Practical Programming in Tcl and Tk.* New Jersey: Prentice-Hall, 1995.

White, J. "A Common Agent Platform," position paper for the *Joint WWW Consortium/OMG Workshop on Distributed Objects and Mobile Code,* Boston, Mass., June 1996.

White, J. E. "A High-Level Framework for Network-Based Resource Sharing, (1996): 45: 561–570.

White, J.E. "Telescript Technology: The Foundation for the Electronic Marketplace," General Magic White Paper, General Magic, Inc., 1994.

White, J.E. "Telescript Technology: Scenes From the Electronic Marketplace," General Magic White Paper, General Magic Inc., 1995.

Wu, Y., "Advanced Algorithms of Information Organization and Retrieval," Thayer School of Engineering, Dartmouth College, 1995.

Zayas, E., "Attacking the Process Migration Bottleneck," *Proceedings of the 11th ACM Symposium on Operating Systems Principles,* (1987): 13–24.

index

A

J

L

M

N

O

P

R

The mobile agents CD-ROM

The CD-ROM that comes with *Mobile Agents* contains all of the software you need to build and run mobile agents! In addition, the CD-ROM contains HTML pages that explain how to install the software agents, plus more information about mobile agent software, papers, conferences, and Web sites. For complete information on the various mobile agents on the disk, please refer to the `README.TXT`, which can be opened in any standard text editor.

Specific mobile agent systems included on the CD-ROM are:

Telescript Telescript was designed by James White at General Magic and is the best example of a complete mobile agent system. The complete Telescript system along with extended documentation and links to General Magic's newest intitiatives are included.

Agents for Remote Access Ara was designed by Holger Peine at the University of Kaiserslautern and was predicated on the belief that users should be allowed to write mobile agents in various existing programming languages. Ara agents are presented in multiple languages and all work in a similar fashion.

Agent Tcl Agent Tcl was designed by Robert Gray at Dartmouth College and presents an agent architecture similar to Telescript, but based on an open language, Tcl.

Aglets Workbench Aglets Workbench was designed by a team at IBM Tokyo and is the most complete agent solution (at the time of publication) based on Java.

More mobile agent information

The limitations of writing a book about bleeding edge software have been overcome through the use of extensive HTML pages on the CD-ROM. There simply wasn't enough time to write about all of the work that is being completed in the area of mobile agents, or other agent technology. Consequently, the HTML pages contain collections of URLs which point to research groups and companies working in these areas. Pointers to discussion groups and Web sites containing more agent information are also included.

The Web site for the book is `http://www.manning.com/Cockayne` and will contain information about new editions of the book as well as URLs for updated software and examples.